RARE
Poultry Breeds
David Scrivener

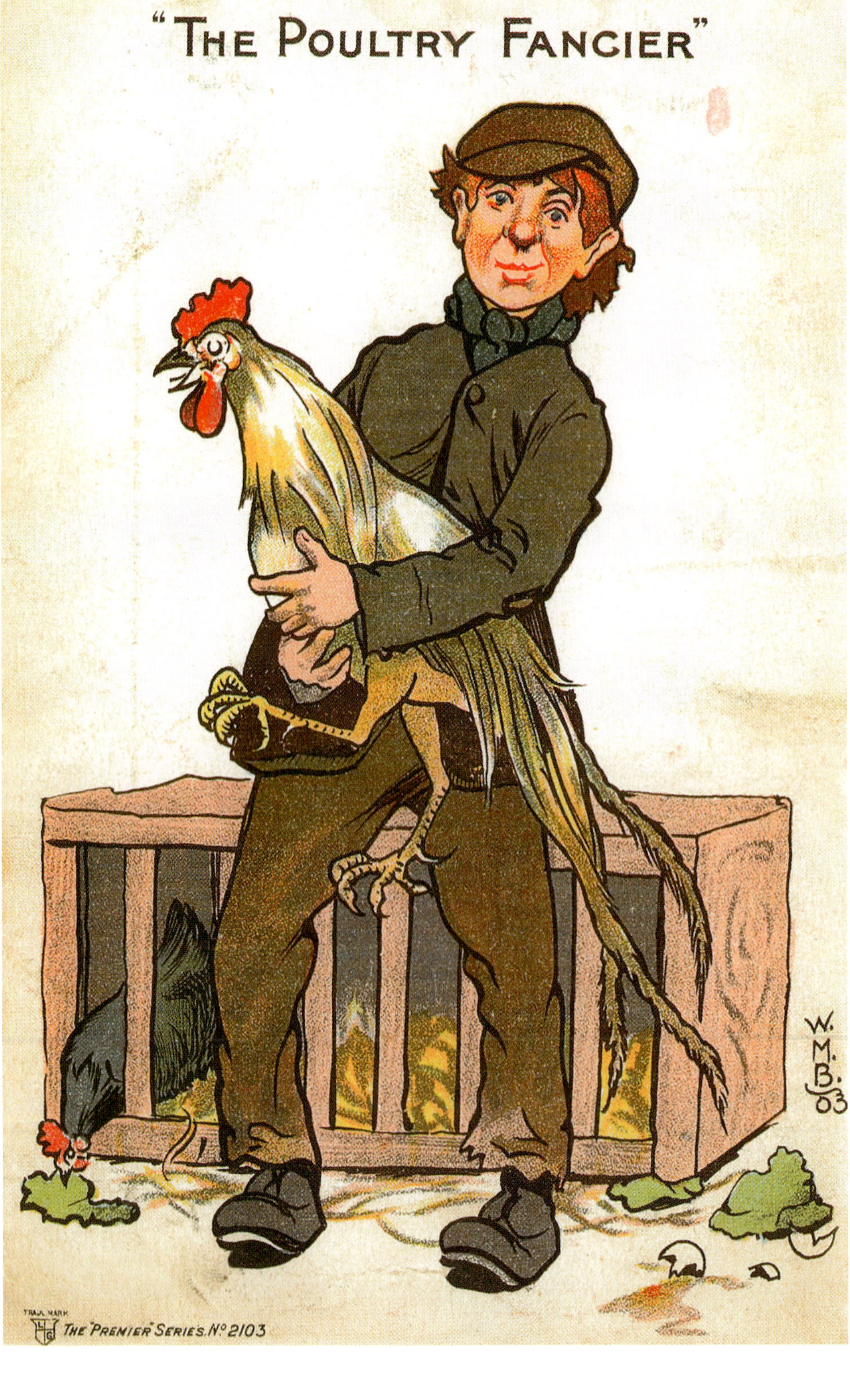

"THE POULTRY FANCIER"

RARE
Poultry Breeds

David Scrivener

THE CROWOOD PRESS

First published in 2006 by
The Crowood Press Ltd
Ramsbury, Marlborough
Wiltshire SN8 2HR

www.crowood.com

© David Scrivener 2006

All rights reserved. No part of this publication may be reproduced or transmitted in any form or by any means, electronic or mechanical, including photocopy, recording, or any information storage and retrieval system, without permission in writing from the publishers.

British Library Cataloguing-in-Publication Data
A catalogue record for this book is available from the British Library.

ISBN 1 86126 889 0
EAN 978 1 86126 889 1

Acknowledgements
Thanks to: Penny Hawtin for the use of some of the cards in her collection; Julia Keeling for the photographs of her trip to Japan; Hans Schippers for the many photographs of European breeds; Mrs Joyce Tarren for allowing me to buy her late husband's (John Tarren) collection of photographs; Fred Hams, Ian Kay and Andrew Sheppy for help with details of some breed histories, including the use of the 'Bucks County Fowl' picture from Fred's collection. Most of the old artist-drawn pictures are from the author's collection.

A note on measurements
Breed weights are quoted in metric or imperial measures, according to the practice of the country of origin of the breed in question.

Typeset by Textype, Cambridge

Printed and bound in Singapore by Craft Print International Ltd

Contents

Introduction	6
1. European Light Breeds	9
2. Egyptian Breeds	79
3. Long-Tailed and Game-Related Breeds	83
4. Rumpless Breeds	107
5. Long-Crowing Breeds	115
6. Crested Breeds and Their Relations	123
7. Cup-Combed Breeds	155
8. Frizzled and Naked-Necked Breeds	163
9. Short-Legged Breeds	175
10. Langshans	181
11. European Medium/Heavy Breeds	199
12. American and Canadian Breeds	219
13. True Bantams	235
14. Autosexing Breeds	251
Bibliography	266
Index	268

Introduction

Everyone interested in old poultry breeds, show exhibitors and others, will have some reference books, the number depending on how bookish they are, how long they have been interested and how lucky they have been in finding some of them. There were several major books with wonderful colour plates; those by W.B. Tegetmeier, Harrison Weir and Lewis Wright are recognized collectors' items. Perhaps less obviously collectable, but actually much rarer, are some of the modest paperbacks and flimsy leaflets, very few of which have survived a century or more. *Rare Poultry Breeds* includes a lot of information that is otherwise very difficult to find.

There have, of course, been other comparable books published in recent years, and this is a good place to mention some of them. British poultry judge Ian Kay wrote *Stairway to the Breeds* (1997), and perhaps best of all (but in Germany) was the excellent pair of books by Rüdiger Wandelt and Josef Wolters: *Handbuch Der Hühnerrassen* (1996) and *Handbuch Der Zwerghuhnrassen* (1998). It would have been nice, and a challenge, to write (virtually) an English version, covering all known breeds, of Wandelt and Wolters pair of books; but even if possible, it would have been too large, and therefore too expensive to publish.

The more popular breeds have been extensively written about elsewhere, both in the breeds chapters of general poultry books and in specialized single breed books. Therefore it seemed a good idea to leave all the popular breeds out, and write in depth about the rarer breeds that are given scant attention, in fact often not even mentioned, in most poultry books.

This is essentially a history book, and the birds described here are living history. For many decades it has been possible to divide fanciers into three groups: first, competitive exhibitors, those likely to keep one of the popular show breeds, Sebrights or Wyandottes perhaps. Second, those who keep pure breeds for 'self-sufficiency', including many Araucana and Welsummer breeders. Third, those who keep 'Rare Breeds' mainly because they are ancient and rare. Some very new breeds, the Eikenbruger from Germany and the Serama from Malaysia, have been omitted because, in the author's opinion, they are too new. As this book is primarily intended to be used by hobbyist breeders, some commercial breeds have been omitted, especially a number from the former Soviet bloc, details of which are in Wandelt and Wolters. Another group of breeds that are omitted (apart from Rumpless Game) are the Game breeds. These have also been widely written about elsewhere, and if they had been included, would have required a lot of details about cockfighting, that even in a historical context would have been a controversial topic. *Rare Poultry Breeds* is not a book on 'how to keep chickens' – for that see the author's other book *Exhibition Poultry Keeping* also published by The Crowood Press.

This leaves every page available to describe and give the history of the breeds

Introduction

'The car's made a top-hole run!', a postcard circa 1930. Artist: unknown

covered in as much detail as possible. Where known, precise details of the creators of breeds, including their addresses if available, have been given. Sample show entry numbers have also been given, partly to give an indication of their popularity in their heyday, and partly to show that there was a heyday. In some cases, Modern Langshans for example, there were days when more birds were on display in one show hall than the entire world population today.

As this book was written in the UK, the definition of 'Rare' has naturally been from a UK perspective. Therefore, all the breeds covered by the British 'Rare Poultry Society' have been included except Belgian Game and Spanish Game, both excluded because they are game breeds. Some foreign readers might think some choices strange, no shortage of German Langshans in Germany for example, but very few anywhere else. Redcaps, Lincolnshire Buffs, Scots Dumpies and Scots Greys, none of which are officially classed as 'Rare Breeds' in Britain, are included here because they are all very rare worldwide. Many of the breeds described here do not exist in the UK at all. Those that have been included are the breeds that are known to be well established in their country of origin, and as a result are likely to be recognized if ever imported here, or to any other country where they have never existed before.

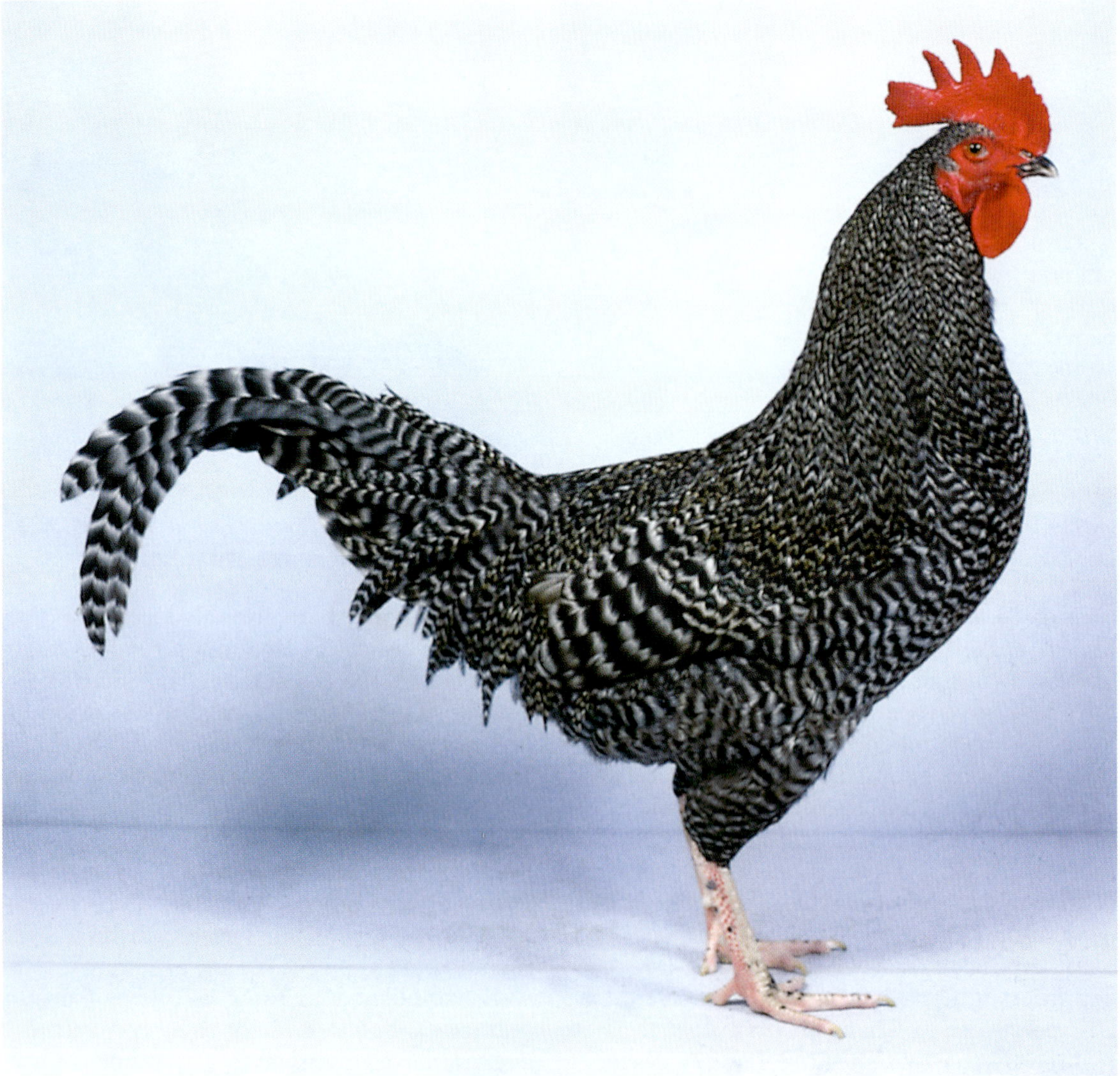

CHAPTER 1

European Light Breeds

The many breeds in this category have been described, alphabetically, in one very large chapter because the complex history of some of them did not allow a sensible division into separate chapters for 'French Breeds', 'Belgian Breeds', and so on.

Breeds not included in this book because they are too popular are: Ancona, Hamburgh, Leghorn (all national types, including Italiener), Minorca, Poland.

The breeds covered in other chapters of this book are:

Autosexing Breeds: Cambar, Crested/Cream Legbar, Legbar, Welbar
Crested Breeds: Altsteirer, Appenzeller Spitzhauben, Brabançonne, Brabanter, Crèvecoeur, Houdan, Sulmtaler, Sultan, Yorkshire Hornet and related La Flèche and Nederlandse Uilbaarden
Short-Legged Breeds: Courtes-Pattes, Krüper, Scots Dumpy
Long-Crowers: Bergischer Kräher, Bosnian Crower
Rumpless Breeds: Ardennaise sans queue/Ardenner Bolstaart

Breeds in this chapter, and their countries of origin:

L'Alsacienne – France (related to German Rheinlander)
Andalusian – Spanish ancestral stock made into a standard breed in UK
Appenzeller Barthuhner – Switzerland
Ardennaise/Ardenner Hoen – Belgium
Assendelftse Hoen – Netherlands
Bergischer Schlotterkamm – Germany
Brakel – Belgium
Bresse-Gauloise – France
Campine – Belgium (Hen feathered standard form established in UK)
Casteliana Negra (Castilian) – Spain
Catalana del Prat – Spain
Caussade – France
Ceská Slepice – Czech Republic
Chaamse Hoen – Netherlands (related to Belgian Brakel)
Cotentine – France
Dansk Landhøne – Denmark
Derbyshire Redcap – England
Deutscher Sperber – Germany
Drentse Hoen – Netherlands
Empordanesa – Spain
Fammene-Hoen – Belgium
Fries Hoen (Friesian) – Netherlands
Gauloise Dorée – France
Géline De Touraine – France
Gournay – France
Groninger Meeuwe – Netherlands
Hergnies – France (related to Belgian Brakel)
Herve-Hoen – Belgium
Lakenfelder/Lakenvelder – Germany/Netherlands
Landaise – France
Le Mans – France
Limousin ('Coq De Péche') – France
Marsh Daisy – England
Norske Jaerhøn – Norway

Old English Pheasant Fowl – England
Ostfriesische Möwe – Germany
Penedesenca – Spain
Ramelsloher – Germany
Rheinländer – Germany (related to French Alsacienne)
Sachsenhuhn – Germany
Scots Grey – Scotland
Spanish – Ancestral stock from Spain, standard breed developed in Netherlands and England.
Starumse Rondkamm – Netherlands (related to Assendelftse and Derbyshire Redcap)
Thüringer Barthuhn – Germany
Ukrainskaja Uschanka – Ukraine
Vogländer – Germany
Vorwerkhüihner – Germany
Westfälischer Totleger – Germany
Zottegems Hoen – Belgium.

L'ALSACIENNE

The eastern border of Alsace is the River Rhein, with Germany on the other side, so it is not surprising that this breed is effectively the French version of the German Rheinländer. They were recognized as a breed in 1897, with the first major display being at the 1907 Strasbourg Show. They rose to become a popular breed up to 1914, with a breed club forming in 1909. This continued to be the case in the early 1920s, but by 1930 they had declined considerably, not reviving until the 1960s.

L'Alsacienne Naine (Bantam)

These were developed during the 1980s. No details were available at the time of writing, but it assumed that almost all L'Alsacienne Naine breeders are in their home region.

Description

Generally they look like very deep-bodied Hamburghs with an almost horizontal back line when standing normally and with large well-fanned tails. They have low-set rose combs that have slightly higher leaders than the close to the skull (as Wyandottes) leaders on Rheinländers. Ear lobes are of moderate size and white. The most popular colour variety is Black, the other colours being Blue, White and Black-red/Partridge.

Standard weight: Large: cock 3kg (6½lb), cockerel 2.5kg (5½lb), hen 2.5kg (5½lb), pullet 2kg (4¼lb). Bantam: cock 950g (33½oz), cockerel 850g (30oz), hen 850g (30oz), pullet 750g (26oz).

ANDALUSIAN

Leonard Barber is believed to be the first to have imported fowls from Andalusia to England in 1846 or 1847. His came from Jerez de la Frontera, about twenty-five miles from Cadiz. They were all excellent layers of large white eggs, but only a few had blue-grey plumage. John Taylor of Cressy House, Shepherd's Bush, London also imported some in 1850 or 1851, and was credited as being the main promoter of the breed for the first decade. His experiences were described in Wingfield and Johnson's (1853) and Harrison Weir's (1902 and 1905) poultry books. Mr Taylor won prizes at Birmingham, Halifax, Cheltenham, Yarmouth, Chevely and Hitchin in 1852, and at a large show at Baker Street in London in 1853. At the same show, Andalusians were also entered by Edward Simons, J. Whittington and W. Culter.

Mr Taylor obtained his first stock from Mrs Synge, of Glenmore Castle, Dublin, 'of which some were grey, others speckled white and black'. These would now be described as 'Blue' and 'Splash'. He also bought some from Mr Barber, and then asked a friend who worked for the 'Peninsular and Oriental Mail Company' in Spain to look out for more on his travels around the country. His friend must have seen hundreds, if not thousands, of chickens in the farmyards and markets of Spain, but only sent a dozen deemed near enough to the required type back to England. Only one 'grey' (Blue) cock and and a pair of 'white' (Splash) birds were actually used by Mr Taylor for breeding. His description of the

European Light Breeds

type and colour he was looking for seems to be the beginning of the standard Andalusian that all breeders have been trying to achieve ever since. 'Plumage bluish-grey or dove-colour, each feather being lightly margined with a darker tint', suggesting that sharp lacing did not then exist.

Wright's *Book of Poultry* (1872 and later editions) mentioned another significant importation, although one that became significant by accident. They were on a Spanish trader's ship in Portsmouth harbour in 1851, and were initially bought by Mr Richardson, an importer of foreign cage-birds. He bred from them, to build up a quantity to sell. One of his customers was Mr Coles, of Farham, Hants. Both men crossed the original imports with Spanish to obtain white lobes and large Mediterranean-type combs. The originals were said to be more game-like in headpoints. All of these early breeders were then confident that the odd black and white (Splash) chicks that they bred were an indication of previous random matings, and that they would soon breed only Blues. The truth about the genetics of Blue plumage would not be discovered until the research by Bateson and Punnett at Cambridge was published in 1906.

Harrison Weir also mentioned 'the old Cornish and Devon blues, with which they [Andalusians] were undoubtedly crossed'. He did not fully explain what these birds were; it is probable that they were strains of Blue Pit Game. Many of the earlier Andalusians were almost self blue with little or no lacing, although many of these pioneer breeders could

Andalusians. An idealized pair as depicted by Joseph Williamson Ludlow (1840–1916). This picture was originally given as a free gift with Poultry *magazine circa 1912–14.*

European Light Breeds

easily imagine the potential. By the late 1890s some definitely laced Andalusians were being exhibited, but the reality was a long way short of the ideal depicted by the artists.

There was an early breed club, but for reasons the author has not been able to discover, it was a combined 'Leghorn, Plymouth Rock and Andalusian Club'. The exact dates of activity of this club are not known either, but it was already active in 1892, and probably dissolved about 1924. A more conventional single 'Andalusian Club' was formed between 1902 and 1910, and seems to have lasted until 1939. A few enthusiasts kept them going in Britain without a breed club until The Rare Breeds Society was formed in 1969. A new Andalusian Society was formed in 1976, but only lasted until 1981, so Andalusians are still a 'Rare Breed' in the UK.

Since 1950 there have seldom been many shows in the UK with more than ten large Andalusians, plus perhaps a similar number of the bantam version. The old Andalusian Club that appears only to have catered for large fowl, was able to attract many more than this, quite an achievement considering that their favourites were never a serious commercial laying breed and even after the genetics of their plumage colour was understood by the scientists, this information would not have reached many hobbyist poultry keepers.

The (British) Andalusian Club Shows usually consisted of four basic classes (cock, hen, cockerel and pullet), sometimes with additional classes for novices and selling birds. From 1910 until the 1920s this was held in conjunction with a number of northern shows, Sheffield, Leeds, Liverpool, Manchester and Glossop. Entries were usually between forty and sixty, which fell to thirty in 1925, no doubt affected by the general economic depression of the time. Club show entries continued at much the same level, twenty-five to thirty-five, up to the outbreak of the Second World War, with a peak in 1934, where fifty-one appeared at

Andalusian, large male. Photo: John Tarren

Andalusian, large female. Photo: John Tarren

York. Large Andalusians were revived and promoted in the UK by Rex Woods of Ashe House, Devon in the 1970s. He obtained his stock from W. Langton of Scotland, who may have had pre-war stock.

Andalusians were admitted to the American Standard of Perfection in 1874, but they were not very popular. As an indication, the highest

Andalusian, bantam male. Photo: John Tarren  *Andalusian, bantam female. Photo: John Tarren*

show entry the author found when researching for this chapter was at the Madison Square Gardens Show, New York, 31 December 1912 to 4 January 1913, where eight cocks, nine hens, seven cockerels, seven pullets and two trios appeared. John H. Robinson suggested (in *Popular Breeds of Domestic Poultry*, Ohio, USA, 1924) that some American fanciers made their own strains from crosses of Leghorns and other breeds instead of importing stock from England or Spain.

They were first exported to Germany from England in 1872. One of the leading German 'Andalusier' fanciers was Herr Heinrich Marten of Lehrte. A specialist club was formed in 1895 at Schmölln, Thüringen, making it one of the oldest poultry breed clubs in Germany.

Andalusian Bantam

Miniature Andalusians have been bred by a few people with the patience to accept the high 'wastage rate' inevitable when trying to breed just a few with even ground colour, sharp lacing, and correct size, type, comb, lobes and so on. The Countess of Dartmouth was mentioned by Entwisle as being a successful exhibitor with them back in the 1880s. It was recommended by Entwisle, Proud and others that the best way to make Andalusian Bantams was to cross under-sized large Andalusians with over-sized Black Rosecomb Bantams and select future generations from there. Despite looking through a considerable number of British poultry show results from 1892 to the present, no significant entries of Andalusian Bantams were found.

Very few records of American shows were available, but it does not seem likely that these bantams were much more numerous there until the 1990s when a joint breed club for Andalusians and Spanish was formed that attracted some interest. Sixty-two large and fifteen bantam Andalusians, along with twenty-nine large and eighteen bantam Spanish were entered at the 1998 Ohio National Show.

The greatest interest in Andalusian Bantams has been in Germany, where they are called Zwerg-Andalusier. It is believed they were first shown there in 1920, but they were lost during Germany's troubled history of hyper-inflation of the 1930s, followed of course by the Second World War. New strains were made when peace returned in the 1950s, and by the 1990s some of the larger German poultry shows had over seventy Zwerg-Andalusier on display.

European Light Breeds

Description

The general size and shape is typical of Mediterranean breeds, an upright, active bird with a fairly large tail. Their combs are of medium size, upright on males, falling to one side on females. Ear lobes are white and almond shaped; eyes dark-red or brown. The most common fault seen as far as their head-points are concerned is white spreading to the faces. This is often seen on older birds, especially cocks, of all white-lobed breeds, but it has been seen on current-year Andalusians at the shows, which indicates that a serious problem is developing. Legs and feet should be dark slate or black.

British standard weights are overly optimistic, requiring 3.6kg (8lb) large cocks, therefore the more realistic German standard weights are given here: large: male 2.5–3kg (5½–6½lb), female 2–2.5kg (4¼–5½lb). Bantam: male 900g (32oz), female 800g (28oz).

The ideal plumage pattern, a soft, even blue-grey ground colour with sharp black lacing and hackle feathers has been a favourite subject of all the famous poultry artists. The idealized colour plates must have encouraged many fanciers to try them, only to give up when put off by the high wastage rate. In addition to the unshowable blacks and splashes, very few of the Blues have sharp lacing or an even ground colour.

Some Andalusian breeders cull black and splash chicks at day old, but most rear them until they are old enough to sex, and then kill the cockerels (unless very short of stock). The best of the black and splash pullets (for size, type, comb, lobe, and so on) can be mated with blue cockerels, using lightish cockerels with black pullets and darkish cockerels with splash pullets, to build up the breeding flock, which is especially useful if not many blue pullets are available. If there were a clear 'best mating', after over a century, someone would have discovered it by now.

APPENZELLER BARTHUHN

Appenzell is the north-eastern Canton of Switzerland, bordering Germany and Austria. The Barthuhner (bearded hen) breed was first made about 1860 from crosses between Brown Leghorns (Rebhuhnfarbigen Italiener), Polveras (related to bearded Polands/Paduaner) and (possibly) Russian Beardeds (Russkije Borodatye). They almost died out during the Second World War (despite Switzerland being neutral), and were revived in the early 1950s by K. Fischer of Stuttgart, Germany. He used some elderly survivors of original strains, Rheinländers, Salmon Faverolles and Thüringian Beardeds. A miniature version was made during the 1990s, very new compared to most pure breeds of poultry. Most breeders of both sizes of Appenzeller Barthuhner are in Switzerland, Germany and Austria, although a few are in the UK.

Description

Appenzeller Barthuhner are quite thickset, with adult cocks weighing 3kg (6½lb), compared to only 2kg (4¼lb) for the much smaller Appenzeller Spitzhauben cocks. Barthuhner Bantams range from 700g (24oz) pullets up to 1,000g (35oz) cocks. They have rose combs and beards. The two original colour varieties were Black and Black-Red/Partridge, with Blues and Blue-Reds appearing recently.

ARDENNAISE/ARDENNER HOEN

These are a standardized form of ancient cottagers' chickens kept for centuries in the Ardennes region of south-east Belgium. Edward Brown (*Races of Domestic Poultry*, 1906) summed them up precisely: 'The Ardennaise fowl is very hardy when kept under natural conditions, but does not bear confinement well. It is an excellent forager, wandering far in search of nourishment.' He went on to say that they were nervous of strangers, that the hens were fair layers of medium-sized eggs, and the cockerels were meaty for their size but needed to be hung for a few days before cooking. No doubt their active life made the meat tough if cooked immediately after killing.

Their home region includes a lot of moorland, peat-bogs and woodland; which suggests that the people in the area (except for recent decades) would not have been able or willing to spare much valuable grain to feed poultry. There are many other breeds that lay larger eggs and more of them, and that are larger and more tender table birds, but they all need proper feeding and housing. Ardennaise would roost up a tree or on barn roof beams, and spend their days hunting for worms. The chickens and eggs had not cost their owners anything.

Moves to make them into a standard breed started in the 1880s, with a first breed club, the 'Union avicole de Liège' being formed in 1893. This was replaced in 1921 by the 'Groupement Avicole de la Haute Belgique pour la Protection de L'Ardennaise'/'Pluimveevereniging van Haute België voor de Bescherming van net Ardenner hoen'. Very few are bred outside their home region, with the possible exception of neighbouring parts of northern France. There is also a rumpless version, Ardennaise sans queue/Ardenner Bolstaart, which is further described in the chapter on Rumpless Breeds.

Ardennaise Naine/Ardenner Kriel/ Ardennes Bantam

Similar bantams may have been known in the area long before, but concerted efforts to produce a specific Ardennes Bantam started about 1904. Willems and Brandt (*Studie over Hoenderachtigen en Watervogels*, 1971) named Théo de Lame, Er. Humblet, G. Ruwet, J. Marichal, M. Dion, M. Eloy and a Mr. Wodon as the leading pioneers. They started by crossing large Ardennes with Old English Game Bantams and Bassette Liègoise/Luikse Bassette Bantams. They made enough progress to stage an unofficial display at the 1907 Liège/Luikse Show. A breed standard was published in 1913, quickly followed by a proper Club Show on the 25–27 January 1914, also at Liège. Here three trios and forty-three single birds were staged of three colour varieties: 'zilver, goud en citroengeel' (silver, gold and lemon-yellow). It is not entirely clear if these were colour variants of the Black-Red/Partridge/Duckwing pattern or the Birchen/Grey/Brown-Red pattern. Eventually the full range of colours (*see below*) as for the large breed were reproduced as bantams.

Description

They are very much a normal light breed chicken in general shape and size. In the days when most British farmyard flocks were bred from Old English game cocks running with assorted crossbred hens, the resulting birds would have been very similar. They have compact bodies, fairly large fanned tails and legs that are long enough for such active birds, but do not appear particularly 'leggy'. Combs are single, straight, and fairly small.

Standard weights: large cock 2–2.5kg (4¼–5½lb), cockerel 1.75kg (3¾lb), hen 1.75–2kg (3¼–4¼lb), pullet 1.5kg (3¼lb). Bantam male 600–650g (21–23oz), female 550–600g (19–21oz).

Colour varieties: Black

Gold Hackled Black (as Brown-red Modern Game, but solid black breast)
Gold Hackled Black (as Brown-red Modern Game, with laced breast)
Silver Hackled Black (as British Norfolk Grey, solid black breast)
Silver Hackled Black (as Birchen Modern Game, laced breast).

(All the above have dark eyes and dark-red or purple comb, wattles, faces and lobes.)

Gold Partridge (females as Partridge Old English Game)
Gold Partridge (females as Partridge Wyandottes, but pencilling less distinct)
Silver Partridge (females as Silver Duckwing OEG)

Silver Partridge (females as Sil. Penc. Wyandottes, pencilling less distinct)
White.

(All these have brown or red eyes, red comb, lobes and so on, and slate blue shanks and feet.)

Special Requirements
As may be guessed, Ardennes, large or bantam, need large grassed runs with very high fences. Potential show birds will need a lot of handling to tame them.

ASSENDELFTSE HOEN
This breed is from the northern part of the Netherlands, and is closely related to Fries Hoen (Friesian). Edward Brown (*Races of Domestic Poultry*, 1906) went so far as to describe them as 'Rose combed Frieslands'. They are also related to Pencilled Hamburghs, indeed they were exported to England as Pencilled Hamburghs in the early nineteenth century, later more specifically as cock-breeder Pencilled Hamburghs. There were always many more Gold than Silver Assendelftse Hoenders, which is probably why the same is true of Pencilled Hamburghs. There is probably also a connection to Derbyshire Redcaps, as both breeds have very large rose combs, and the early exports to England would explain why British strains of Hamburghs usually had larger combs than Hamburghs in other countries.

Assendelfter Kriel (Assendelfter Bantam)
These are a recent creation, just dating back to the early 1980s. The author has only seen photographs of them, and these suggest they might be written off by some judges as just very bad Pencilled Hamburgh Bantams. More work seems necessary to develop uniformly small bantams with characteristically large combs.

Yellow Pencilled Assendelftse, large female. Photo: Hans Schippers

Description
General body shape is similar to Hamburghs, and they have the same slate-blue shanks and feet and white ear lobes. They have large rose combs, similar to that of present day Derbyshire Redcaps. There are two colour varieties, gold and silver versions of the 'ears of wheat' design well known on Friesians and Sicilian Buttercups. The official name of the golds has been changed to 'Yellow Pencilled' to emphasize that the ground colour should be yellow, not the deep gold of Gold Brakels. Markings remain black, not white as on Yellow Pencilled Friesians.

BERGISCHER SCHLOTTERKAMM
This breed is from western Germany near the Dutch border, especially around Düsseldorf. They are the best of the traditional German breeds as layers of large white eggs. It is believed they were developed about 1800 from crosses between Mediterranean stock (Castilians?) and pre-existing local breeds, such as Bergischer Kräher, Hamburghs and

Thüringian Beardeds. The Mediterranean breeds would help to increase egg size.

Schlotterkamms are standardized in four colour varieties, two of which are conventional Black and Barred, and the other two are almost unique, shared only with Bergischer Krähers. These are called 'Schwarz-Weißgedobbelt' and 'Schwarz-Gelbedobbelt', and are described below. It is not clear how long the Bergischer Kräher breed or this plumage pattern has existed. Pictures of Schlotterkamms dating from the 1900–25 period, show birds that are basically similar, but not exactly the same as current standards.

Bergischer Zwerg-Schlotterkamm (Berg. Schlotterkamm Bantam)

This is a recent creation, having been developed between 1985 and 1990. They were bred from crosses between large Schlotterkamms and Bergische Zwerg-Kräherhner (Bergischer Kräher Bantams).

Description

General shape and size is as one would expect from the breeds used in their make-up. They seem to combine the long-backed style of Minorcas with the compact, meaty, rounded breast of some western European varieties that are good quality table birds despite being light breeds. When standing correctly for a show judge, the back should be horizontal and the top line of their moderately large and fanned tail should be about 45 degrees above the horizontal.

They have medium-sized single combs, upright on males. Hens' combs are upright at the front, with the hind part flopping over. Wattles are of normal size, as also are their oval, white, ear lobes. Eyes are chestnut-red and shanks/feet are slate-blue.

Standard weights: large male, 2–2.75kg (4¼–6lb), female 1.75–2kg (3¾–4¼lb). Bantam male 900g (32oz), female 800g (28oz).

Black and Barred Schlotterkamms do not need much description, and are hardly ever seen anyway. The mottled or 'gedobbelt' pattern does need detailed attention, however. The pattern is the same on the black/white and black/gold varieties. At first glance, the males look like Black-Red (black/gold) or Duckwing (black/white) males except for the wings, that are heavily laced. The females are almost solid black, again except for heavily laced wings. Both sexes have a few white spots on their breasts and elsewhere. Close examination shows that these spots are in the centre of the feathers, not white tips. It is known that spangling (as on Hamburghs) and lacing (as on Laced Polands) are genetically related, as can be seen on some breeds with 'moony' markings of Derbyshire Redcaps, Old English Pheasant Fowl, and the uneven lacing of Brabanters. The pattern of Schlotterkamms and Bergischer Krähers is probably a darker ('melanistic') version of one of these patterns, perhaps based on a different pattern on the 'E locus' than the Columbian pattern of the other Spangled and Laced breeds.

BRAKEL

Brakel is a village near the market town of Alost/Aalst, north-west of Brussels in Belgium. They were kept by almost every farmer and smallholder in the area until serious egg producers switched to hybrids during the 1960s. Fortunately, there is still an active breed club in the area, with an annual club show at Alost/Aalst.

There are several similar breeds nearby. Campines are from the neighbouring area of Belgium to the north-east of the Brakels' region, up to the Dutch border. On the other side of the border, in the Noord-Brabant region of The Netherlands, is the Chaams Hoen. Going south, in the French Département du Nord, are Hergnies fowls. All are described in this chapter.

Prof. Dr A. E. R. Willems and E. D. Brandt (in *Studie over Hoenderachtigen en Watervogels*, 1971) mentioned an article in an 1884 issue of *Avicultura* magazine as the first effort to regard Brakels and Campines as two

Silbergedobbelt Bergischer Schlotterkamm, large male. Artist: August Rixen

Silbergedobbelt Bergischer Schlotterkamm, large female. Artist: August Rixen

separate breeds, and that this was followed in 1888 by a more detailed monograph on the differences between them. At that time the fowls of both regions had the same plumage pattern, with fully cock feathered males, as only standardized for Brakels now. The hen feathered Campine males came later – *see* Campine section for full details. So, back in the 1880s the only difference was that the Brakels were heavier than Campines. This was because there was a more developed egg and table poultry industry around Aalst, the market there being an important buying point for wholesale buyers from Brussels. The Campine ('Kempische') region was heavily wooded with a light sandy soil, and therefore a less prosperous agricultural economy. Campine fowls still expected to forage for their own food and roost up a tree, on the beams of the cowshed, or wherever. The greater care taken in the breeding and management of Brakels resulted in a 'Braekel Club' being formed at Gent/Ghent on 1 May 1898.

It was sheer bad luck that Campines became widely known in Britain, America and elsewhere, wheras Brakels remained virtually unknown to anyone apart from those with a very keen interest in exotic poultry breeds. In 1897, Mr Vander Snickt (a Belgian poultry expert) escorted a group of similar experts from other countries around many poultry breeders and shows. This party included Edward Brown, a British poultry book author, and Mr A. F. Brown, editor of *Farm Poultry* magazine in the USA. They happened to visit Campine breeders, not Brakel breeders, so the Campines got the publicity when they returned home. In 1906, Edward Brown (in *Races of Domestic Poultry*) freely admitted that had he seen Brakels on that trip, he would have favoured them over Campines. He imported some in 1898, but by then it was too late, Campines were already

European Light Breeds

Silver Brakels, pair. Artist: Kurt Zander

established in the minds of British poultry keepers, even though a lot of the 'Campines' imported to the UK were probably Brakels really.

At the time of writing (2005), there were less than ten known Brakel breeders in the UK, the most prominent being Brian Sands of Lincolnshire.

Braekel Naine/Brakel-Kriel/ Brakel Bantams

There have been sporadic attempts at producing these since they were first recognized as a standard breed around 1900. At least as many breeding programmes in this direction have been done by German fanciers as Belgians, a reflection of the greater interest in fancy poultry in Germany. The normal process has been, as with most miniature fowl, to cross the smallest available birds of the large fowl breed in question with whichever bantam breeds seemed most appropriate. Sebrights, Schwarze Deutsche Zwerghühner (Black German Bantams), Hamburgh Bantams and Black Rheinländer Bantams have all been mentioned. Earlier strains were lost in the chaos of two world wars and the almost as bad years in between. They were remade by several fanciers between 1955 and 1975 in Belgium and both East and West Germany. A few flocks have existed in England since 2000, all Silvers.

Description

General body shape and size is similar to other European light breeds, but they are rather more deep-bodied and meaty. The breast is noticeably full, wide and rounded, probably the result of crosses with Dorkings over a century ago to improve their table bird

European Light Breeds

qualities. When standing normally, their back slopes slightly from shoulders to tail, reflecting their traditional dual-purpose function, livelier than specialist table breeds, more sedate than laying breeds or those that were expected to forage for their own food.

Combs are single, upright, medium to slightly larger than medium in size. They have white oval ear lobes, dark brown eyes and slate-blue shanks and feet.

There are two standard colour varities, Gold and Silver, different colour versions of the same pattern. Gold females have black barring over a rich gold ground colour in all parts except for the head and neck that is clear gold. The black bars should be wider than the gold ground colour bars in between. Gold Males are the same except for having clear gold shoulders, back and saddle feathers as well. Not many males have clear barring on their tail feathers. Silvers are the same, with pure white replacing the gold ground colour. In both varieties there is a tendency for birds with the best body feather barring to have a few black neck feathers, but these are normally removed for showing.

Gold Brakel, large male. Photo: John Tarren

Weights, Netherlands standards: large, male 2.25–2.75kg (5–6lb), female 2–2.5kg (4¼–5½lb). Bantam, male 800–900 (28oz–32oz), female 700–800g (24–28oz).

Special Requirements

Brakels need large houses with the perches well away from the house walls, to avoid their long tail feathers being broken. Generous-sized grassed runs should be provided if possible, ideally with some trees or other source of shading for Silvers. Breed as many as possible each year to be sure of getting enough with correct markings.

BRESSE-GAULOISE

This breed, once usually called 'La Bresse', now has the double name 'Bresse-Gauloise' because the specialist breeders in its home area secured an 'Appellation d'Origine' for the name 'Bresse' in 1988. Only birds bred in the area near the city of Bourg-en-Bresse in south-eastern France, between Lyons and the Swiss border can now legally be called Bresse. These regulations are no doubt enforced on commercial producers elsewhere in Europe, but it is not clear if it would be enforced in the show world. To avoid potential legal problems Bresse are now classed as additional colour varieties of the Gauloise breed. 'Gauloise Doree', traditionally only bred in the Black-Red/Partrdge colour variety, are covered later in this chapter. The original colours of Bresse, Black, White and 'Grey', despite now being (to adapt a famous phrase) 'The breed that dare not speak its name', are described here.

In the 1923 *Poultry* magazine yearbook Mrs Onslow Piercy said the 'records in France prove that the breed existed in 1591'. In the 1930 edition of *Feathered World Yearbook* Gerald Fox similarly affirmed that Bresse date back to the sixteenth century. Neither author gave any details of their sources. Moving on, Harrison Weir (in 1905) said that Bresses were mentioned in a French treatise on poultry in 1810. Lewis Wright

European Light Breeds

Silver Brakel, large male. Photo: John Tarren

Silver Brakel, large female. Photo: John Tarren

Silver Brakel, bantam male. Photo: John Tarren

quoted Edward Brown saying that Bresse fowls were first noticed by the British between 1815 and 1818. 'Noticed' is almost certainly a euphemism for 'looted', as the British in question were the troops occupying Bourg-en-Bresse after the Napoleonic Wars. As peace returned to France, and transportation improved during the nineteenth century, Bresse fowls were bred in ever larger numbers to supply not only Lyons and other French towns and cities, but also markets in Switzerland and northern Italy.

The Bresse breed may be an amalgamation (before the recent Gauloise issue) of three much more ancient, and more localized strains. Edward Brown said that Black Bresses were originally, Varieté de Louhans', Grey Bresses were 'Varieté de Bourg' and White Bresses were 'Varieté de Beny-Marbez'. Blacks were said to be the best layers and Whites the best table birds. A special system of fattening was used, involving forced feeding of damp mash down their gullets, to transform these light breed cockerels into plump roasters. 'Poules de la Bresse' were featured on restaurant menus all over France and beyond. Bresse cockerals are very meaty for their size and are fine-boned.

Edward Brown imported some Bresse to England in 1895. They may not have been the first ever Bresse in the UK, but this importation started the brief flurry of interest in the breed here up to their eventual decline during the 1930s, and extinction by 1945. The author does not know if there are any in the UK at present – that is, 2005.

The 'British Bresse Club' (sometimes called 'English Bresse Club') was formed in 1911. From 1926 to 1928 there was a rival 'White Bresse Club' which favoured higher weights for the British table bird market, but they amalgamated into the 'British United Bresse Club' in 1928. No definite information was available to the author, but it is assumed that the club was suspended in 1939, and was never revived after 1945.

In the first few years the British Bresse Club had entries of: thirty-nine birds in 1911, fifty-two in 1921 and fifty-two again in 1913 (eight White M., ten White F., eighteen Black M., sixteen Black F.) at their annual club shows. There was one amazing display of over 150 birds at the White Bresse Club show at Tottenham, London in 1926, but this was marred by a lot of the birds being dirty, with none of the preparation expected by judges and regular show exhibitors. The only other large display in Britain was the 1931 Bresse Club Show, where 100 appeared, mostly Whites, a fair number of Blacks and a few Greys. Despite all the leading breeders here being well aware of their status as one of the original colour varieties in France, Grey Bresse were never standardized in Britain. Bresse breeders here were a utility-minded lot, did not enter shows very much, and when they did, during the 1930s, competed in the 'Utility Section'.

No references to Bresse Bantams have been found. There are a number of broadly similar light breed bantams in existence already, so it is very doubtful if they would be worth making.

Description

At first glance Bresses are similar to most other breeds in this chapter in general body shape, but when handled they are found to be noticeably meatier. The breast should be well rounded and deep, with a fairly long and broad back. Tails should be moderately well developed and carried at an angle of about 45 degrees.

They have medium-sized single combs, upright on males, flopping over on females. Ear lobes are white, eyes black or dark brown, shanks and feet are blue-grey on all varieties. Beak colour varies from dark horn on Blacks, through shades of blue to bluish-white on Whites. The facial skin is red or sooty. Standard weights range from 2kg (4¼lb) pullets up to 3kg (6½lb) adult cocks.

Blacks and Whites are the two main varieties, Blues and Buffs were also mentioned in the 1920s and 1930s. 'Grey' Bresse are a

White and Black Bresse, pairs. Artist: A.J. Simpson. This picture was originally a free gift with The Feathered World *magazine, circa 1925–30.*

lightly marked variation of the Silver Pencilled pattern of Hamburghs, or to be more exact, Silver Assendelfters or Silver Groninger Meeuwe.

CAMPINE

In their home region of northern Belgium, up to the border with The Netherlands, these were simply village hens, very similar to Chaams Hoen north of the border and the more selectively bred Brakels further south in Belgium. They only became a refined exhibition breed when in the hands of British enthusiasts. It was these British breeders who established hen-feathered males as the norm, Belgian breeders had usually killed them as 'unnatural', preferring to keep normal cock-feathered males, like Brakels for breeding.

Campines were brought to the attention of British breeders by Edward Brown after his tour of the Belgian poultry scene in June 1897. One of the first importers was Mr Thomas B. Bracken of Slyne Road, Lancaster, who brought some over in late 1897 or 1898 from M. Moons de Coen of Calmpthout, Antwerp. Mr Bracken was successfully showing them at Crystal Palace and elsewhere by 1900. A British Campine Club was formed at one of the major shows in 1900 or 1901. Show entry figures are often a good guide to the broader popularity of any breed; in this case, the creditable display of seventy-two Silvers and twenty Golds at the 1901 Alexandra Palace Show, London, indicates that Campines were off to a good start.

The first indication that the British breeders would go for hen-feathered males can be found as quoted from Mr Bracken in Lewis Wright's *New Book of Poultry* (1902):

> In order to avoid that bane of so many otherwise good breeds, double mating, The Campine Club have recently revised their standard, and made the plumage of cocks identical with that of hens; and, with a view to discourage the introduction of Hamburgh blood, have also made red eyes a disqualification.

Although some Campine Club members had seen hen-feathered males, none of them seem to have had any. A leaflet published by Campine Club secretary, Rev. E. Lewis Jones, said the first cockerel fitting the revised standard was a Silver which won the major awards at the 1904 Alexandra Palace Show. It was exhibited by J. Wilson of Penrith and was from eggs sent over by Oscar Thomas of Renaix, Belgium. Many sons of this bird were sold to fanciers around the country, making this one of the cases where a whole breed can be traced back to one bird.

This development temporarily made Gold cocks, still all cock-feathered, unshowable. Rev. Lewis Jones and Mr E. S. Mattock set to work, crossing hen-feathered Silver males with Gold females. By 1911 there was a total of five hen-feathered Gold males in existence, and as happened with the Silvers, their sons were spread around the country.

Rosecombed Campines briefly appeared, between 1915 and 1925. Richard Edwards had a rosecombed Silver cockerel chick hatch in 1915, bred from his own champion strain of hens and a cockerel bought in from another reputable breeder, W.G. Reiss. More were bred from this cockerel once it reached maturity, and they were distributed to other fanciers. However, good these rosecombed Campines were in other respects, some were said to have had excellent markings, most judges simply 'passed' them at the shows because rose combs were non-standard. Special classes were put on for them at a few shows in 1921, followed by a short lived separate Rosecombed Campine Club, circa 1922–24, after which they were finally accepted by the main Campine Club. Although rose combs were still included as an option in British Standards until the 1971 edition, it is doubtful if any have existed since the 1930s.

The British Campine Club continued to 1969, when it merged with the then newly formed Rare Breeds Society (later Rare Poultry Society). The last Club Show entry details found by the author were for 1962, when there were thirteen Silver males, fifteen Silver females and seven Golds in a mixed M/F class. It is doubtful if there have been any larger dispays since then.

Campines were a favourite with American poultry artists such as Louis A. Stahmer, but only enjoyed a brief boom in real life. Arthur D. Murphy of Maine is believed to have been the first importer, circa 1893. He tried to popularize them, had them included in American Standards, but failed, and they were even dropped from the Standards in 1898, and not reinstated until 1914. Mr R. Jacobus of Ridgefield, New Jersey imported some Silver Campines from England in 1907, which presumably had better markings than Mr Murphy's Belgian stock. This importation attracted enough interest to warrant the formation of the American Campine Club on 22 December 1911 at the Madison Square Gardens Show, New York. In 1912, it had 177 members, but it was not to last, White Leghorns swept every other breed away for commercial egg production during the 1920s. In this early period, 1907 to 1920, Silvers were much more popular in the USA than Golds. Only Arnold G. Dillsbury of Pennsylvania (who also kept Silvers) had Golds good enough to win 'Champion Campine' at the major shows over Silvers.

Silver Campines, pair. Artist: J.W. Ludlow. This picture was originally a free gift with Poultry *magazine, circa 1912–14.*

Campine Bantam

No references have been found for these before 1950, and neither have any explanations for their absence, considering that most other established breeds were bantamized long before 1940. Since the 1950s, several attempts were made to produce Campine Bantams in a number of countries. The author does not know if any exist now, and if so, where they are.

Description

General body shape and size is similar to other European light breeds. Because of their current low numbers and inevitable inbreeding problems, size and vigour must be carefully maintained. The hen feathering of Campine males also needs constant selection to keep as it should be – many have some silvery, Brakel type, saddle feathers. If there are a lot, the bird should be culled. Just a few such feathers, and it has long been normal practice to remove them for showing.

Campines have medium-sized single combs, upright on males, falling over on females. There was once a rosecombed variety, but most people thought them too much like Hamburghs for them to have widespread credibility. Ear lobes are oval, white and of medium size. Campines have dark brown eyes, an important breed characteristic. This is sometimes associated with dark facial skin,

European Light Breeds

Gold and Silver Campines, pairs. Artist: A.J. Simpson. This picture was originally a free gift with The Feathered World *magazine, 5 July 1912.*

although bright red faces are required. Shanks and feet are 'leaden blue'.

(British) Standard weights: large, male, 2.7kg (6lb), female, 2.25kg (5lb). Bantam, male 680g (24oz), female, 570g (20oz).

Plumage pattern and colours: for full details of the pattern, consult American or British Standards. In brief, Golds have clear, rich gold neck feathers with the remaining parts barred, black and gold. Silvers are the same pattern, black and white. 'Buff' or 'Chamois' Campines, as Golds but with white replacing black, were mentioned (1924) but never standardized.

Before being 'prepared for exhibition', most good Campines have a few black neck feathers that are removed. Birds with naturally clear necks seldom, if ever, have strong enough barring. The part most likely to fail is the upper chest and up between the (clear) neck feathers. The black bars should be two or three times as wide as the gold or silver bars in between. If the barring of each feather coincides, giving a 'ringing' effect, this is considered an extra point of quality. It is very rare for the barring to be perfect throughout, tail feathers being most likely to fail in this respect.

Silver Campine, large male. Photo: John Tarren

Silver Campine, large female. Photo: John Tarren

CASTELIANA NEGRA/ CASTILIAN

In his *Races of Domestic Poultry* (1906), Edward Brown said he saw a display of Castilians at the 1902 Madrid Show. They were an attempt 'to rejuvenate the native fowls' from which Andalusians, Minorcas and White-Faced Spanish had been developed. As a standard breed, only the Black variety is recognized, but there would probably have been other colours when they were 'native fowls'.

Edward Brown said that they were sometimes called 'Moorish Fowl', possibly indicating that they were first brought to Spain when parts of it were under Islamic rule from about AD 700 until 1400. He also mentioned another alternative name, 'Zamorana Fowl', from the name of the region around Madrid.

Present-day Castilians are essentially the product of German fanciers. Wandelt and Wolters (1996) mentioned Dr H. Zolke, who brought hatching eggs from Spain in 1955, and developed a uniform black strain from them. They were standardized in Germany in 1957; and have a distinctive 'look', difficult to describe, but you know it when you see it, distinguishing them from other black-plumaged, white-lobed, Mediterranean fowls.

Castilian Bantam

These are a recent creation, work starting in the 1970s, also in Germany. Large Castilians were crossed with Black Dutch Bantams, Black Minorca Bantams and Black Rheinlander Bantams. As far as the author is aware (in 2005), Castilian Bantams are only bred in Germany, and are not very popular there.

Description

As said above, Castilians are a typical Mediterranean breed in general type and size. The comb, of single type, is upright on males, falling over on females, and of moderate size. Ear lobes are oval, white, and no more than 'medium' in size. Eyes are red-brown.

Standard weights: large, male 2–2.5kg (4¼–5½lb), female 1.75–2kg (3¾–4¼lb). Bantam, male 850g (30oz), female 750g (26oz).

CATALANA DEL PRAT

These are from the Prat district, in the Cataluña region of Spain, near Barcelona. They were bred from Buff Cochin crosses from about 1870 onwards and, as a result, are larger than other Spanish breeds. Catalanas can be regarded as the Spanish equivalent of Lincolnshire Buffs and prototype Buff Orpingtons or Rhode Island Reds. The main local breed they were crossed with was said (ref. W. Powell-Owen, 1932 FWYB) to be the 'Muriana', which must have been a very local breed, as the author has not found any other references to them. Powell-Owen gave a clue about Murianas when he said that one of their characteristics was a double-ended or side-sprigged single comb that was also often seen on Catalanas well into the 1920s. Some Catalanas even had crests, as seen in a photo in Ian Kay's *Stairway to the Breeds*. Single combs with large side sprigs are characteristic of two other Catalañ breeds, Empordanesas and Penedesencas, both of which are described later in this chapter. Divided single combs also suggest a relationship with Sicilian Buttercups.

Once they were stabilized only Black-Tailed Buffs were standardized, but as would be expected, there was a lot of variation in the early days. Dr A.E. Shaw was said to have perfected the 'Buff' variety (Black Tailed Buff) and Señor B. M. Barreto concentrated on the short lived Partridge variety. These were a light partridge, and as far as one can see from a grainy old black and white photo of a hen, seem to have been similar to Quail Barbu d'Anvers in colour.

Edward Brown (*Races of Domestic Poultry*, 1906) saw them in 1902, where they were first exhibited at the Madrid World Fair. Being a very new breed, they were still very variable, some having residual leg feathering. This was naturally the first time they were seen by international poultry experts. Then they seemed likely to become a medium or heavy breed, but over the intervening century they settled back as a white-lobed breed,

Buff Catalana del Prat, large female. Photo: Hans Schippers

meatier than typical Mediterranean breeds, but just about still in the Light Breed category.

Professor Salvador Castelló, Director of the Spanish Royal School of Poultry Breeding at Arenys De Mar, Barcelona, took some to South America, possibly in 1914 when he went on to rediscover Araucanas in Chile. Catalanas became very popular in several South American countries, particularly Argentina. W. Powell-Owen saw them there on a trip in the late 1920s or early 1930s, saying that they were suited to the large estancias, being active foragers, laying almost as well as White Leghorns, in addition to the fact that Catalanas were much better table birds. Some Argentinian Catalanas were exhibited at the 1930 World Poultry Congress at Crystal Palace, London, but if they *were* left here (cheaper than taking them back), they did not catch on.

Possibly because of their following in South America, Catalanas were admitted to American Standards in 1949, but they did not catch on there either. There was just one Catalana at the 1998 US National Show at Ohio.

Catalana Bantam

These were admitted to American Standards in 1960, but only as suggested weights with the note: 'Shape and color descriptions the same as for large Buff Catalanas.' No records have been found of them ever existing, as is also the case for several other bantam versions of large breeds. Bantam weights were given as a guide should anyone make them in the future.

Description

General body shape and size is between the light breeds and the medium-weight dual-purpose breeds, perhaps similar to the lighter weight and larger combed specimens of Welsummers. The breast should be broad, deep and rounded; and the back long and sloping slightly down to the tail when standing normally.

The comb is single and moderately large, straight and upright on males; and upright at the front on females, with the rear part falling over to one side. Eyes are reddish-brown, ear lobes white, oval and although described as 'large' in American Standards, would be described as quite modest by British fanciers. Shanks and feet are slate-blue. Plumage is a rich golden-buff with a black tail. *See* the American Standard of Perfection for full details.

Standard weights: large: American and German Standards are slightly different, but roughly range from 5lb (2.25kg) pullets up to 8lb (3.6kg) adult cocks. Bantam: range from 24oz (680g) pullets up to 32oz (910g) adult cocks in American Standards, but these seem unrealistically low for a bantam which, if ever bred, would be expected to have utility value.

CAUSSADE

Apart from having shorter legs, Caussade fowls are similar to Black Bresse-Gauloise. They are from the Languedoc-Roussillon region, to the south-west of the Bresse fowl's home area. Like Bresse; Caussades are good layers of white-shelled eggs, and the cockerels are good quality, if rather small, table birds. Cocks weigh about 2kg (4¼lb), and hens 1.5kg (3¼lb). There is only one colour variety, black. Ear lobes are white, and combs are single and upright on males and pullets, flopping over on adult hens. They have bluish-grey shanks and feet. The author had no definite information, but believed at the time of writing that very few Caussades are bred outside their home region.

CAUX

This was once the main utility breed of the Normandy region, and were recorded as far back as the 1770s. They continued to be bred right up to 1939 but the entire breed seems to have been a casualty of the Second World War, although it is possible that a flock exists, unknown to the author somewhere. If extinct, there may not be enough difference between Caux and Black La Bresse for a remake to be viable.

CESKÁ SLEPICE

This breed from the present Czech Republic appears to be a reincarnation of the Magyar breed described by Edward Brown in 1906. Some readers may need to be reminded that the Czech Republic was once part of the old (pre-1918) Austro-Hungarian Empire. Czechoslovakia (now divided into the Czech Republic and Slovakia) was created as an independent country in 1918.

This is a compact and meaty-looking breed, but at weights ranging from 1–8kg (4lb) pullets up to 2.5kg (5½lb) adult cocks are quite small. Apart from having red ear lobes they are very similar to Drentse Hoen (white lobes) from the Netherlands. The main colour varieties are gold and silver versions of the Black-Red/Asiatic Partridge (as Wyandottes) pattern, gold and silver versions of the Pencilled (as Hamburghs?) pattern, Black, White and Black-Mottled (as Houdans). The author did not know at the time of writing if

European Light Breeds

this breed is bred outside its homeland.

CHAAMSE HOEN

This is, or was, the Dutch equivalent of the Belgian Brakel. Chaams fowl were widely bred in the Noord-Brabant region of The Netherlands in the late nineteenth century and formally standardized in 1911. The author is not certain, but believes Chaams fowl were extinct by 1945, with a revival from Brakel crosses being made in the 1980s.

The plumage pattern is as Brakels, mostly as Silvers. Chaams fowls are required to have orange eyes instead of the dark eyes of Brakels and Campines.

COTENTINE

A black plumaged light breed from the Normandy region of France. Weights range from 2kg (4¼lb) pullets up to 3.5kg (7¾lb) adult cocks. It is not known if they are still bred in their home region; they are almost certainly unknown anywhere else.

Gold Pencilled Chaamse, head of male. Note red eye colour. Photo: Hans Schippers

Silver Pencilled Chaamse, male. Photo: Hans Schippers

Silver Pencilled Chaase, female. Photo: Hans Schippers

DANSK LANDHØNE (DANISH LANDHEN)

The beginning of poultry showing in the mid-nineteenth century encouraged enthusiasts to draw up breed standards for local fowls all over the world. Danish poultry fanciers refined their native chickens into a standard breed between 1875 and 1905. Very few Danish Landhens are bred outside Denmark.

They are a compact light breed, with weights ranging from 1.75kg (3¾lb) pullets up to 2.5kg (5½lb) adult cocks. Combs are single and upright, lobes are white and fairly small, and shanks and feet are slate-blue. The main plumage colour variety is Black Red/Partridge (females are of a similar colour to Welsummers). Other varieties are Silver Partridge (Silver Duckwing), Barred, Black, Buff and White.

DERBYSHIRE REDCAP

There is a Derbyshire Redcap Club in Britain, which means they are not officially classed as a 'Rare Breed' in the UK, despite being very rare indeed. Most of the club members live in the midland counties of England, probably with no more than 500 birds between them. Redcaps are virtually unknown outside the UK.

Before organized poultry shows started in the mid-nineteenth century there were informal poultry competitions, usually held in pubs. In northern England, some of the ancestors of Hamburghs were shown under a variety of local names, including Bolton Bays, Bolton Greys, Lancashire Moonies and Yorkshire Pheasants. Derbyshire Redcaps were another such local breed, but along with Old English Pheasant Fowl, have retained their separate identity up to the present day. Early editions of Wright's *Book of Poultry*, from 1872, mentioned that 'years ago' there were special classes at shows in the Sheffield area for Redcaps, but that they had died out by the time of writing (1870?). 'They appeared to be a kind of mongrel Golden-spangled [Hamburgh], larger in size, and with *immensely* large rose combs hanging over at the sides.'

No details were found by the author about any early clubs for Redcaps, but there certainly was a Redcap Club 'in a very flourishing condition' by 1911. It held a Club Show at Manchester that year in which twenty-two Redcaps (eleven F, eleven M) were entered. The secretary of the club, Joseph Heathcote, of Birchover, near Matlock, Derbyshire, won 3rd F, 1st and 3rd M. In the next two years, entries rose and then stabilized, with thirty-four (nineteen F, fifteen M) in 1912 and thirty-three (eighteen F, fifteen M) in 1913. Until then there had not been any Redcap classes at the other 'classic' shows, London Dairy, Birmingham and Crystal Palace, but two members guaranteed classes for the 1913 Crystal Palace, which then had six males and nine females entered. Show organizers then were cautious about the classes provided because of the generous prize money. 'Guaranteeing' a class meant agreeing to pay all or part of the prize money if there was not enough entries, and hence entry fees, to cover this expense. Local interest continued after the First World War, with club membership rising from fifteen in 1920 to fifty-three in 1921. The Club Show at Buxton on 11th December 1920 had thirty-two birds entered.

The breed standard was drastically revised in 1922 to reduce the required comb size of cocks. For 'some twenty years' (since 1902?) the top surface of the comb on a mature cock was expected to be 5½in long by 4½in wide (140 × 115mm) at its widest point. As a result many combs lopped over to one side and the sheer weight of the comb was injurious to their health. This was reduced to 3¼ × 2¾in (82 × 70mm), I believe, with judges instructed to make allowances for a couple of years for everyone to adjust.

This proved to be a wise move by the leading breeders, most of which no doubt knew all there was to know about managing giant combed Redcaps. The 1923 Redcap Club

European Light Breeds

Derbyshire Redcaps, pair. Artist: J.W. Ludlow, circa 1890. These birds were 'Bred by, and the Property of, Mr Albert E. Wragg, Edensor, Bakewell.

Show at Bakewell attracted a record ninety-eight entries. By 1925 interest was even growing out of their home area, confirmed by two well-filled classes at Barnstaple Show in Devon. In 1926 there was probably the largest ever entry of Redcaps at a classic show well away from the northern or midland counties – i.e. forty-six at Crystal Palace (eight cocks, eleven hens, sixteen cockerels, eleven pullets) A slow decline set in, but there were still some impressive displays to be seen, seventy-five at the 1928 Club Show at Bonsall (near Matlock, Derbyshire).

1930 saw the World's Poultry Congress at

Derbyshire Redcaps, pair. Artist: J.W. Ludlow. This picture was originally a free gift with Poultry *magazine, 4 April 1913.*

Crystal Palace, the most important few days for UK poultry keepers during the inter-war years. As far as possible, every breed was to be represented by the best available birds from each participating country. Redcaps from Derbyshire could be seen alongside others from Canada and the USA. These had changed somewhat since the original stock had been taken across the Atlantic, with a lighter plumage ground colour and combs with smaller 'workings'.

In the 1933 *Feathered World Yearbook*, Mr W.H. Baker, who had been keeping Redcaps for over fifty years, said that over half the breeders lived within ten miles of Bakewell, although the remainder were well spread around the country. He went on to describe how they were one of very few breeds who could survive, let alone thrive, in the 'cold, bleak district with long winters and short summers, (that) frequently attains an altitude of over 1,000ft above sea-level, where wheat refuses to ripen and apples to mature'. Redcaps were never used commercially on a large scale, but remained popular on smallholdings through to 1945, but then rapidly declined. By 1960 there were probably less than ten flocks, and the club had closed down.

Interest in old pure breeds revived in the late 1960s and early 1970s. Mr Syd Brassington and Mr Fred Kitchen got the Derbyshire Redcap Club going again in 1976. As before, a lot of the forty or fifty members the club usually has are in Derbyshire and surrounding counties. At major shows, circa 2005, expect to see twenty

European Light Breeds

or more entered. Interest is currently rising, so a display of up to forty birds is not impossible.

Derbyshire Redcap Bantam

Redcap Bantams were first developed around 1930 by Messrs Baker, John Elliott and H.O. Young. They started from crosses between undersized large Redcaps and Partridge Wyandotte Bantams. As nearly white ear lobes (should be red on Redcaps) have been a persistent problem, they were certainly not going to use Hamburgh Bantams. It is possible that Old English Game Bantams were used as well, but no definite mention of this has been found. Mr Underwood of Nottingham exhibited some around 1950, but it is not certain if this was a new strain or from the 1930s stock. The author started to regularly attend poultry shows from about 1970, but did not see any Redcap Bantams until the 1990s, and then only a few. They are attractive birds, and it would only take a few keen breeders to transform their fortunes from the dire estimated 2002 total UK breeding females population of less than twenty hens.

Description

In the show hall, headpoints are the main priority, given 45 out of 100 in the Scale of Points. Even at the reduced standard sizes since the 1920s, Redcap combs are still impressive. The top surface of their combs should be covered in 'workings', or fleshy spikes. They are longer and larger than any other rosecombed breeds seen at British shows. In the Netherlands they are replaced by Assendelftse Hoenders and Starumse Rondkam that have similar combs. One noticeable difference between Redcaps and all its relations – that is Assendelftses, Hamburghs and Old English Pheasant Fowls, is that Redcaps have red ear lobes and all the others have white. Wattles are moderately large, the beak is horn colour and eyes are red. A tendency to white lobes and undersized combs has been seen, probably a result of some crossing with Old English Pheasant Fowl. Some of this crossing may have happened by accident, as these two breeds look broadly similar. It is probably not a good idea for anyone to keep both breeds.

General body shape and style is robust and muscular, but Redcaps are still a light breed. They have normally been allowed extensive free-range conditions, and need them to keep fit. As they have slate blue shanks/feet and brown and black plumage this does not present any problems for exhibitors.

Normal weights are: large males 2.5–3kg (5½–6½lb.), large females 2–2.5kg (4½–5½lb). No Redcap Bantam weights published. 700g (24oz) pullets up to 1000g (35oz) cocks seem realistic.

Full details of the plumage pattern are given in *British Poultry Standards*, but it can be summarized as a mix of glossy nut-brown and black. The males are mostly black, with the glossy brown being limited to parts of the neck, back and saddle. Females are mainly rich brown with half-moon black spangling. This has been the only colour variety in the history of the breed, and any attempts at additional colours are likely to get a hostile reception from the traditionalists.

DEUTSCHE SPERBER

'Sperber' is German for Sparrowhawk, a species that has breast plumage that appears to be barred. German poultry breeders have used the word 'Sperber' in place of 'Cuckoo' as used by their British counterparts. Otto Trielhoff of Duisburg was the originator, who initially thought of calling them 'Rheinischer Sperber'. Breeds he used to make included: Barred Italiener (= utility Leghorns), Barred Plymouth Rocks, Bergische Schlotterkammen and Black Minorcas. He started about 1901, and circa 1907 almost decided to go along the lines of Barred Minorcas instead. It is not known why this path was not pursued, possibly because the other German Minorca breeders would not accept them. Progress seems to have come to a standstill until 1917, and as Germany was losing the First World War at the time, there would not have been a

Derbyshire Redcap, large male. Photo: John Tarren

Derbyshire Redcap, large female. Photo: John Tarren

cross with Scots Greys. The breed still exists in Germany, but is even rare there. They are unlikely to attract much interest in the UK, being too similar to our native Scots Greys.

Deutscher Zwerg-Sperber/Deutscher Sperber Bantam

These were made in the 1960s by Paul Adam of Mülheim/Ruhr and Heinz Rehme of Osnabrück from large Deutscher Sperbers and Deutsche Zwerghennen (German bantams). Barred Zwerg-Italiener (± = Leghorn Bantams) were introduced a little later. Probably because of the Deutsche Zwerghuhners, Deutscher Sperber Bantams have longer tails (in proportion) than the large fowl version.

Description

General shape and type is very similar to American style Minorcas. They have moderate sized single combs, upright on males, and upright at the front, drooping to one side after the first point, on females. Ear lobes are white and of a similar size to those of American Minorcas, that is, smaller than UK Minorcas. They have white skin, shanks and feet, and the barring is fairly coarse. Tails are fanned, fairly large on the large fowl, larger in proportion on the bantams.

DRENTSE HOEN

These are known to have been bred in the Drenthe region in the north-east of The Netherlands since the seventeenth century. No detailed historical information was available to the author until Edward Brown's *Races of Domestic Poultry* (1905) that mentions a Drentse cock depicted in a piece of needlework dated 1805. This region was quite isolated and rural until modern transport made us all part of 'the global village'. Most poultry were kept 'free range' in the real traditional sense of the phrase, that is, they had to forage for their own food. They are one of the smaller light breeds, as might be expected from their history.

When interest in poultry farming and

lot of progress then. However this is the time when the name 'Deutsche Sperber' was introduced. Around this time another breeder, A. Schneider of Dresden, tried a

European Light Breeds

Deutscher Sperber, large male. Artist: August Rixen

Deutscher Sperber, large female. Artist: August Rixen

showing grew in the late nineteenth century, Drentse were neglected. The heavy Asiatic breeds were bringing changes to poultry keeping in The Netherlands as well, leading to the creation of Barnevelders and Welsummers among others. By 1900, Drentse were almost extinct, but they were saved by Roel Houwink (1869–1945) from Meppel. He gathered as many surviving specimens of the breed as he could find to revive and stabilize a range of colour varieties. Before then, they were a very mixed bag.

They were threatened again by the Second World War, and this time were saved at the government poultry research institute ('Rijksinstituut van de Pluimveeteelt'), 'Het Speldeholt' at Beekbergen. So many were bred at the institute that they were even kept commercially for a few years after the war. Drentse are now a popular hobbyists' breed in The Netherlands, but at the time of writing (2005), the author was not aware of any in the UK.

Drentse Bolstaart/Rumpless Drentse

These may have existed for centuries, but no details have been found. *See* the Rumpless Breeds chapter for local relations, where it will be seen that the nearby Ardennes fowl also has tailed and rumpless versions.

Drentse Hoenkriel/Drentse Bantam

A fairly new breed, made during the 1950s by R.W. de Boer of Wamsveld. They were produced from crosses of large Drentse, Dutch Bantams, Friesian Bantams and local crossbred bantams. Rumpless Drentse Bantams were then made from tailed Drentse Bantams, large Rumpless Drentse and Dutch Bantams.

Description

They are a small breed, but meaty for their size with a well-rounded breast. Plumage is abundant, with flowing neck hackles and long, well fanned tails. Combs are single, medium-sized, and upright on both sexes; ear

European Light Breeds

Partridge Drentse, idealized large pair. Artist: Cornelis Simon Theodorus Van Gink (1890–1968). Van Gink produced many hundreds of fine watercolours and pen drawings of pigeon and poultry breeds during his long career, wrote several books, was editor-in-chief of Avicultura *magazine, and was a leading figure in the World Poultry Science Association.*

lobes are white. Large fowl weights are no more than some breeds of bantam, ranging from 1.3kg (2¾lb) pullets up to 1.9kg (4lb) adult cocks. The equivalent bantam weights are 600g (21oz) and 800g (28oz).

There is a huge range of colour varieties, the most notable being several versions of Black-Red male pattern males with Asiatic type Partridge/Pencilled females, which is normally associated with heavy breeds such as Brahmas, Cochins and Wyandottes; not little light breeds. Five colour versions of this pattern are recognized: Gold (as Partridge Cochins), Silver (as Dark Brahmas or Silver Pencilled Wyandottes), Blue-Gold, Blue-Silver and Yellow (Gold Duckwing males and dilute Gold females). In addition to these, there are five similar coloured males, but this time with females of the pattern one would expect on Modern or Old English Game. Also with game-type patterns, are Creles, Piles, Birchens, Brown-Reds, Lemon-Blues and Silver-Blues. To add to the range, there are Barreds, Blacks, Blues, Whites, Black-Motttes and Red-Mottles/Millefleurs.

EMPORDANESA

These originate from the Cataluña region of north-eastern Spain, and are related to another breed from the same area, the Penedescenca. Both lay dark-brown shelled eggs,

37

European Light Breeds

extremely dark in the case of Penedesencas.

Empordanesas were first made about 1920 by Professor Rosell i Vilà at Empordà and others at the Spanish Poultry Institute at Barcelona. They seem to have vanished for many decades, only to be rediscovered in 1982. No details have been found about who kept them going, especially during the difficult years of the Spanish Civil War (1936–39).

In general appearance, they look similar to Catalana del Prat, but Catalanas do not lay brown eggs, so there must be some signficant difference in their make-up. Therefore they are typical Mediterranean-type fowls, perhaps a shade heavier than some others. There are three plumage colour varieties, Black-Tailed Buff, Blue-Tailed Buff and White-Tailed Buff. They have red ear lobes, as one would expect on a brown egg-laying breed; unlike the unique Penedesenca, the only white-lobed breed that lays dark brown

Patridge Drentse, large female. Photo: Hans Schippers

BELOW: *Partridge Drentse, large male. Photo: Hans Schippers*

eggs. One characteristic shared by Empordanesas and Penedesencas is a standard type of comb that would bring instant disqualification to any other breed at a show. These are single combs with mandatory side sprigs on both sides. It may be in the form of a thickened mass at the rear, or one distinct, large sprig each side. Genetically, they are probably related to cup combs.

Although some Penedesenca have been in the UK since about 2000, no Empordanesas are here as far as the author is aware.

FAMENNE-HOEN

This is a rare Belgian breed. It is a white-plumaged light breed, probably related to White Ardenner-Hoen. They are believed to date back to the mid-nineteenth century, and are from Mache-en-Famenne a town in the Luxembourg province of Belgium. Edward Brown did not mention them; if he saw any he probably assumed that they were White Ardenners.

Famenne-Kriel/Famenne Bantam

These were first made circa 1921–25. The author did not know at the time of writing if they continue to exist as a separate breed, being very similar to White Ardenner Bantams.

FRIES HOEN/FRIESIAN

Bones of small chickens that were probably very similar to Friesians today, have been discovered by archaeologists in the Friesland region of The Netherlands that have been dated as over 1,000 years old. This area is fairly bleak and windswept, and the people who live there are very religious (Calvanist Protestant), have a local dialect that is virtually a separate language, said to be nearer to ancient English than modern Dutch. All these factors have helped to maintain the separate identity of Friesian fowls.

Before poultry showing became established, Friesians were simply the local chickens. They did not have a breed name until about 1880. This may have also been true of Ardenner and Drentse Hoenders. All three were cottagers' fowls who were small and active out of necessity to survive. They had to find most of their own food, escape foxes, and perch up in the rafters of barns or wherever they could.

Once established as a breed, a breed club was formed, the 'Fryske Hinneklub' in Friesian dialect, 'Friese Hoen Club' in standard Dutch. The club had over 300 members in 1995, confirmation of their continued popularity.

During the first great period of pure poultry breeds, when showing was established, and pure breeds were commercial farm chickens as well, say from about 1870 until 1939, Friesians were virtually unknown outside The Netherlands. If mentioned at all, they only appeared under the heading of 'Lesser Known Foreign Breeds' in English and American books, and only a few of these. There are probably more Friesians being bred outside The Netherlands now than ever in previous centuries. Although there are some fifteen colour varieties, almost all Friesians kept outside The Netherlands are the 'Geelwitpel' (standard Dutch), 'Wyt-weiten' (Fries dialect), 'Chamois Pencilled' (English) colour variety.

The Penciled pattern in Friesians is nearer to the Sicillian Buttercup pattern rather than the Pencilled Hamburgh pattern that might be assumed at first glance. It has been bred in several colour combinations, and it is the buff with white pencilling combination that has attracted most attention because it is almost unique, there being a few Pencilled Hamburghs and Brakels in the same colour combination but, obviously, with slightly different markings.

Because Friesians are so similar to Ardenner and Drentse Hoenders, plumage patterns have been used to help preserve their separate identities. Both other breeds have a range of colour combinations of the Black-Red/Partridge pattern, that are there-

European Light Breeds

fore not accepted in Friesians. In Friesians, apart from several colour versions of the Pencilled pattern, there are Cuckoos, several self colours and several mottled (spotted) colour combinations. More detailed descriptions are given below.

Friese Hoenkriel/Friesian Bantam

Friesians are a small breed. 'Large' Friesians are not much larger than most Hamburgh Bantams shown in the UK, which everyone, except UK Hamburgh Club members, agrees are much too big. Consequently, since Friesians have been bred in the UK (since about 1995), large Friesians have frequently been entered in 'Rare Breed Bantam' classes at shows. Novice exhibitors can readily be forgiven for this error, but (unforgivable!) some judges have not known they were in the wrong class and have given them prizes. Real Friesian Bantams are not much larger than Dutch Bantams, a breed that has enjoyed a meteoric rise in popularity. This perhaps bodes well for the future of Friesian Bantams as well.

Description

General shape is typical of active light breeds, with large well-spread tails. The breast is carried high, which with their long wings gives them a style roughly similar to smaller breeds of Pit Game. They have medium-sized single combs, upright on males, almost upright on females. Large Friesian weights range from 1.2kg (2¾lb) pullets up to

Silver Pencilled Friesians, male and five females. Artist: Cornelis S. Th. Van Gink

European Light Breeds

1.6kg (3½lb) adult cocks. The equivalent range in the miniatures is 450g (16oz) to 650g (22oz).

All colour varieties of Friesians have dark orange eyes; small, white, oval ear lobes; and all except the Cuckoo variety have slate blue beak, shanks and feet (Cuckoos – white). The Pencilled pattern, 'Pel' in standard Dutch and 'Weiten' (Wheaten – meaning grains of wheat, not wheaten colour as OEG) in Friesian dialect, is similar to the Sicilian Buttercup pattern. Males have almost clear (gold, silver or buff) plumage except for wing markings and tail colour. The tail is almost solid black on Golds and Silvers, ideally with gold or silver edging on sickles and side hangers. On Chamois Pencilled/Geelwitpel these tail feathers should be white with buff edging. Of the birds seen by the author in the UK, very few had complete edging, a situation that one hopes will improve as the breed is bred in greater numbers. Females have clear gold, silver or buff near feathering, with the same ground colour elsewhere with black or white markings in the form of two rows of elongated spots up each feather arranged as grains on an ear of wheat. During the 1990s two more colour combinations of this pattern have been made, Gold/Blue Pencilled and Silver/Blue Pencilled. Other colour varieties are self Black, Blue and White; Cuckoo, Black Mottled and Red Mottled. The Red Mottled is a modified version of the well-known Millefleur. Red Mottled males (if C.S. Th van Gink's painting was accurate) should be predominantly red with white spots over

Red Mottled Friesians, male and four females. Artist: Cornelis S. Th. Van Gink

41

European Light Breeds

all except for the tail that is black with white tips. There is some black spotting (as Millefluers), but this is limited. The females are more like Belgian Millefleurs, but they also have less black than Belgians.

GAULOISE-DORÉE

This breed is popularly thought of as being the symbolic 'Gallic Cock' of France. Many purists claim, probably with good reason, that this honour should really go to the 'Combattant du Nord', the French fighting cock, political correctness and artists who were not poultry experts being responsible for this deviation from historical accuracy. Artists usually depict the 'classic cock' as a single combed bird of typical light breed type with a large flowing tail in the Black-Red plumage pattern. This is exactly the description of a Gauloise-Dorée cock. Hens are, as one would expect, of the partridge pattern. They have small white ear lobes and slate-blue shanks and feet. Standard weights are 2.5kg (5½lb) males and 2kg (4½lb) females. No references to a bantam version have been found, and not many to the large version either, suggesting there may be more pictures of them than live specimens.

Chamoise Pencilled Friesian, large female. Photo: John Tarren

GÉLINE DE TOURAINE

This breed is (or was, the author is not sure if they still exist) a brown egg layer, at the top end of the weights in the 'Light Breed' category. They were bred from crosses between existing local farmyard chickens and Langshans imported during the 1875–90 period. The resulting new breed was standardized in 1909. A specialist club was formed, the 'Club Avicole de Touraine'; it is not known when, but it was certainly active in the 1920s. Most of the breeders, including the Club President (M.J.B. Martin in 1928) and Secretary (M. Glaser-Wery in 1928) came from the city of Tours in the Département Indre-et-Loire.

Géline de Touraine are (or were) a conventional light/medium-shaped breed with glossy black plumage – exactly what one would expect from Langshan crosses. They have (had) small single combs, lobes with white centres and red edges, and light shanks and feet. Weights ranged from 2.5kg (4½lb) pullets up to 3.5kg (7½lb) adult cocks.

LEFT: *Chamois Pencilled Friesian, large male. Photo: John Tarren*

European Light Breeds

GOURNEY

This French breed originated at Gourney-en-Bray in the Département Seine-Maritime, Normandy. This town is between Rouen and Beauvais. They started from crosses of local farmyard chickens with Brahmas and Houdans made in the mid-nineteenth century. After initial popularity, they soon declined, only to be revived by an enthusiastic group of breeders in about 1896. Presumably, this is when the 'Club de la Gourney' was formed.

Gourney fowls are a compact, plump-bodied light breed. Not very large (1.8kg (4lb) pullets up to 2.8kg (6¼lb) adult cocks), but very meaty. They could be considered as Normandy's equivalent to the famous la Bresse from south-eastern France. Only one plumage colour variety has ever been recognized, an irregular mixture of black and white like Exchequer Leghorns. This was the original colour of Houdans, and still is in France; different from the tidy white spots seen on exhibition Houdans in the UK, USA and elsewhere. Their skin is white, as one would expect on a French table bread, and the shanks/feet are bluish-white with a few black spots. The hens were said to be 'splendid layers of large white eggs'. There is a photograph of a hen in Wandelt and Wolter's 1996 book, so they clearly still exist. No references to Gourney Bantams have been found.

GRONINGER MEEUWE

Groningen is the north-easternmost province of The Netherlands, with the province of Friesland to its west, and the German province of Ostfriesland to its east. It is therefore not surprising that the Groninger Meeuwe breed of chicken is related and very similar to Friese Hoen and Ostfriesische Möwe, the neighbouring breeds. For many centuries they were all one population, only being divided into three breeds in the early twentieth century. Before shows, and breeds standards, they were simply the local hens. According to E. Verhoef and A. Rijs (*The Complete Encyclopedia of Chickens*, 2004), a breeder in the Groningen area about 1900 exhibited some birds that were much heavier than the usual Friese Hoenders (Friesians). From these, Groninger Meeuwen and Ostfriesische Möwen were established as separate, standardized breeds. Groninger Meeuwen were lost, probably during the Second World War, and the current strains date back to the 1970s and 1980s, remade by fanciers from Friese Hoenders and Ostfriessische Möwen, both of which survived the War.

As far as the author (in 2005) is aware, Groninger Meeuwen are only bred in The Netherlands. German fanciers will naturally keep their own Ostfriesische Möwen, and there are enough other broadly similar breeds established internationally.

Groninger Meeuwkriel/Groninger Meeuw Bantams

Miniatures Groninger Meeuwen were made by Dutch fanciers circa 1955–65, who were almost certainly working in close co-operation with German friends just across the border who were making Ostfriesische Zwerg-Möwen at the same time. They remain as localized as the large version, for the same reasons.

Description

These are typical European Light Breed, clean-legged, with large fanned tails, white ear lobes and straight single combs. Their standard weights, and resulting body shape, is halfway between Friese Hoenders and Ostfriesische Möwen. Large males are 1.8–2.1kg (4–4½lb), large females 1.6–1.8kg (3½–4lb), bantam males 550–650g (19oz–23oz) and bantam females 450–550g (16oz–19oz). If they are ever imported to the UK, breeders and judges should note that the large fowl, although heavier than the Friesians seen in Britain, are still small, and the bantams are the same size as Rosecombs.

Shanks and feet are slate-grey and eyes should be dark brown. The latter point is

European Light Breeds

*Silver Pencilled Groninger Meeuw, large male.
Photo: Hans Schippers*

*Silver Pencilled Groninger Meeuw, large female.
Photo: Hans Schippers*

important, as it provides one of the few definite differences between Groninger Meeuwen and Ostifriesische Möwen that have lighter, orange-brown eyes. Only two colour varieties have existed for most of the last century, Gold and Silver Pencilled. The author did not know at the time of writing if the new Gold/Blue Pencilled, Silver/Blue Pencilled and Buff/White Pencilled varieties of Ostfriesische Möwen in Germany have been accepted in The Netherlands on Groninger Meeuwen. The pattern of the males is much the same as on Pencilled Friesians and Hamburghs, but the females are obviously different. Their neck and upper breast is clear, with very few markings, similar to Friesians, but different from Hamburghs. The remaining markings are larger and coarser than either breed.

HERGINES

Edward Brown (*Race of Domestic Poultry*, 1905) briefly mentioned them, but had not actually seen any. He referred to local expert, M. Lemoine. Hergines are very similar to Belgian Brakels, and have been called the Frnech equivalent to them. Hergines is a village to the west of Mons, very close to the border with Belgium. Before both breeds were bred to precise show standards, both types (many would be hard pressed to spot the difference) were no doubt seen on both sides of the border. Hergines died out, probably during the Second World War, and were remade circa 1987–94 by fanciers B. Dupas, M. Adolphi and P. Kamp.

They are slightly heavier than Brakels. Hergines range from 2kg (4½lb) pullets to 3kg (6½lb) cocks, with Brakels 250g (8¾oz) below for both. Another difference is in facial skin and eye colour, which is why the three fanciers above had to do some 'remaking'. Hergines have darkish facial skin and the irises of the eyes are an unusual mixture of orange and black.

HERVE-HOEN

No information was found when researching this book on their approximate date of origin as an indentifiable breed, but they were mentioned in 1905 by Edward Brown, and so probably date back to at least 1880. All expert authors agree that they were from eastern Belgium in the regions of Limburg and Liège. Brown said they were good layers by the standards of 1900, but the eggs were small. This is not surprising, considering the typical body weights then: 1.4–2kg (3–4½lb) males and 1.15–1.6kg (2½–3½lb) females. Selective breeding and better nutrition, housing, and so on, increased these weights to 2–2.5kg (4½–5½lb) males and 1.5–2kg (3¼–4½lb) females by the 1930s.

In their original form at least, they were hardy birds, active foragers, reliable sitters and 'kloeks', the charming Dutch/Flemish word for mother hens. As they were then a very small breed, they were not great table birds, although some were force-fed to be 'poulets de lait' (milk chickens).

The regional poultry club, 'Union Avicole de Liège' had twenty-four Herve-Hoenders entered in their 1895 show, and published a breed standard on Christmas Day 1896, which they then put up for recognition by the 'Société Nationale pour l'Amélioration de l'Aviculture en Belgique'. They survived the destruction of the the First World War surprisingly well, as indicated by an entry of 149 Herve-Hoenders at the 1923 Liège Show. They remained a popular farm laying breed during the inter-war years in their home regions, despite the increasing domination of White Leghorns, Rhode Island Reds and other 'international' breeds elsewhere, where poultry farming was beginning to change into the poultry industry we have today. However, the world had changed too much in the post-war years, and they rapidly declined. As seen in the description below, this breed is not distinctive enough in appearance to become popular with hobbyists.

In 1989, the Europa Show was in Ghent (it is a huge Show, held every other year, in a different European country each time). Not a single Herve-Hoen was entered. The author did not know (in 2005) if any existed at all.

Herve Kriel/Herve Bantam

These were first made circa 1945–55, just when large Herves were declining. With no interesting colours, patterns or other features, it is not surprising that they did not attract many fanciers away from whatever other breeds they were already keeping. They have survived, as confirmed by two males and four females, all Blacks, appearing at the 1989 Europa Show.

Description

This is a typical single-combed, large-tailed, light breed. They are most likely to be seen at shows in Belgium, where without the relevant paperwork, judges might have trouble distinguishing them from Ardenner Hoenders. Herves have deeper bodies, more fanned tails, more horizontal body carriage, higher wing carriage and larger combs.

There are three colour varieties, each with a traditional name, no doubt associated with a place where some leading breeders once lived. Originally only Blacks were called 'Herve', the Blues were 'Mauheid' and the Cuckoos were 'Cotte de Fer'.

Blacks have a black beak, slate shanks and feet and dark brown to black eyes. Blues have a horn-coloured beak, slate shanks and feet and dark brown (but never black) eyes. The blue plumage has black tips. Cuckoos have a white beak, shanks and feet and orange eyes.

LAKENVELDER/LAKENFELDER

This breed was created in the areas around Utrecht in The Netherlands and in the Nordrhein-Westfalen region of Germany. Lakenvelder with a 'v' is the Dutch version of the breed name, Lakenfelder with an 'f', the German version. They have been bred on slightly different lines for the last century. They have been called 'Jerusalem Fowls', probably as a reference to their black and white plumage being reminiscent of some types of clerical garb.

A Dutch gentleman saw these distinctively coloured chickens at the hamlet of Lakervelt, between Lexmond and Meerkerk, south of Utrecht in 1927. The breed's name is also said to be derived from a white linen sheet ('Laken') over a black field ('Feld' or 'Veld'). Herr Wirz, a customs official from Halden, was the first recorded German breeder, in 1835. It is not known how this difficult to breed plumage pattern was preserved so long ago over a fairly large area for an extended period. This was long before regular poultry shows, and no references to private competitions (as in England with Hamburghs) have been found. The villagers might have favoured this colour combination because they were the same as Belted Lakenvelder cattle, German Pasture/Hannover-Braunschweig pigs (as Saddlebacks) and Dutch rabbits.

Genetically, the pattern seems to be an extremely dark form of the Columbian pattern. Although said to be related to Brakels and Campines, the rare Belgian Zottegems fowl seems to be a closer relation, as also might be 'Moorkop' Brabanters or Nederlandse Uilebaarden, or Black-Crested White Polish.

Herr Rektor Bodelmann of Melle, Hannover was the leading German exhibitor during the 1860s. Lakenvelders remained a popular breed until replaced by Barnevelders and other brown egg layers at the end of the nineteenth century. Curiously, they became better known to the outside world from a large entry at the 1899 International Show at St Petersburg, Russia.

The first known importation of Lakenvelders to England was in 1901, first appearing in public in June 1902 at Shrewsbury Show. A British Lakenfelder Club was formed sometime between 1902 and 1904. Mr P. W. Thorniley of Wem, Shropshire was its first secretary, and as Wem is very close to Shrewsbury, it can be assumed that he was the first importer and exhibitor. Mr W. Bradbury of Ipswich, Suffolk took over as secretary in 1913, but the club ceased to exist sometime during the 1914–18 war. Apart from them not being the most productive breed for a population coping with war-time shortages, the fact that they partly came from Germany would not have helped their popularity here. The Lakenfelder Club does not seem to have been very large or active during its brief existence, only one class of nine birds at the 1912 Crystal Palace Show (W. Bradbury 1st, 3rd (H. Baker 2nd) having been found in the records available to the author. No doubt there were more classes

Lakenvelder, large male. Photo: John Tarren

Lakenvelder, large female. Photo: John Tarren

than this, and Lakenfelders did do some winning in AOV classes, such as a cock winning 1st AOV at 1908 Crystal Palace.

A few Lakenvelders were exported to the USA, sometime between 1900 and 1908. The author has not found any mention of an American Lakenvelder Club, but despite this, has found at least one show with more Lakenvelders than the nine mentioned above. At the Madison Square Garden Show, New York, 31 December 1912 to 4 January 1913, there were four cocks, four hens, four cockerels, four pullets and two trios.

From the 1920s up to the present (2005) there have been a few Lakenvelder breeders in the UK, these few preferring the Dutch spelling 'Lakenvelder'. This is because they have followed the Dutch colour standard (*see below*) that is slightly different from the German. Since 1969 Lakenvelders have been covered by the Rare Poultry Society here, but seldom have more than ten birds been seen at a show. They are much more popular in their home countries, with up to 100 specimens to be seen at the German National Show.

Lakenveld Kriel/Zwerg-Lakenfelder/ Lakenvelder Bantam

The first miniature Lakenvelder was made by a Dutch breeder, Mr Sheeper, during the 1930s, but they died out during the Second World War. They were not seen again until a new strain was made by German breeder, Herr Günter Consbruch of Hamburg in the early 1970s. He used under-sized large Lakenfelders, Vorwek Bantams (that had been made by 1963), Black and 'Hell' (Columbian/Light) Deutsche Zwerghuhner. A new colour variety, with blue replacing black, was made in The Netherlands in the early 1990s.

Description

Size and body shape is much the same as most other European Light breeds, with Lakenfelder tails being larger and more fanned than most. They have upright single combs and small white ear lobes.

Lakenvelder, bantam male. Photo: John Tarren

The general idea of the plumage pattern is the same in all countries, a solid black neck and tail, with a white body in between. There are black markings on the main wing feathers and the saddle hackles of the males. These are heavier in Dutch Lakenvelders than German Lakenfelders. The undercolour is also darker on Dutch birds than German. Very few birds have a perfect separation of black and white, with the tendency for Dutch Lakenvelders to have some black in what should be the white parts, and German Lakenfelders often have white in what should be the black parts. It is considered normal with both types to remove feathers to improve show birds; although many specimens have too many black or white feathers in the wrong places for this to be viable. Juvenile plumage is very variable, and not a reliable guide to their adult feathering. Large numbers have to be reared annually to be sure of some good ones.

LANDAISE

Compared to some other breeds in this book, Landaise are relatively new, having been made just after the First World War, and first exhibited in 1922 at Paris. They are from the Département of Landes. The breeds used to make them were La Bresse, Gasconne and Caussade. Landaise are (or were? – it is not known if they still exist) a single-combed, black-plumaged breed with yellow beak, shanks and feet.

LE MANS/DU MANS

When writing *Races of Domestic Poultry* (1905), Edward Brown said that records of their history and origin were difficult to find. He also said that Le Mans were not bred anywhere else in the world outside their home region, the Département of La Sarthe. This is probably still true today, a century later. In their heyday, despite being a light breed, they were mainly bred for meat, having very good conformation, a lot of meat, with fine bones and comparatively little offal. They were seldom exhibited, even at the major Paris shows, but once dead and plucked, they were exported as far away as Moscow. The original breed died out; the current strain being a remake (circa 1990) from La Flèche, L'Alsacienne and Belgian Game.

Le Mans are a black-plumaged breed with white lobes and skin, black or slate shanks and feet and a large rose comb. Weights range from 2kg (4½lb) pullets up to 3kg (6½lb) adult cocks.

LIMOUSIN

They are also called the 'Coq De Péche', the cock's neck, back and saddle hackle feathers being highly sought-after for making fly-fishing lures. Their place of origin is the Départment of Corrèze, south of Limoges, in the 'Massif Central' of south-central France. The plumage colour is Blue-Red. Limousin have a single comb and dark brown eyes. Weights range from 2kg (4½lb) pullets up to 3kg (6½lb) adult cocks.

MARSH DAISY

Mr John Wright of Marshside, Southport, Lancashire started the process in 1880 that eventually led to the creation of Marsh Daisies. He had two flocks of crossbred chickens: a pen of 'Rosecombed White Leghorns' (the rose combs coming from an earlier cross with a Black Hamburgh cock); and a 'broodies' pen, for hatching pheasant eggs that had been pure Old English Game Bantams until he increased their size with a Malay (large? bantam?) cross. He gave an OEG × Malay cockerel to his friend Mr Wignell, who put it with his White Leghorn hens, presumably to breed some meatier cockerels. Mr Wignell gave one of the cockerels bred back to Mr Wright, who mated it with his 'Rosecombed White Leghorn' hens. This gave good results, so he maintained a closed flock from them for the next thirty years. The hens of this flock came in three colours, Partridge, Wheaten and White.

Mr Charles Moore of Hatfield Woodhouse near Doncaster (two addresses mentioned in the same village were found, 'Lowgrounds' and Beech Trees Farm; it was not known when he was at either residence) came into the process in September 1913, when he bought two of the last of Mr Wright's hens, who by this time was nearly eighty years old, and had not bred any chicks for three years. Mr Moore mated these two hens with a Pit-Game Cock, then with a cockerel from this mating back to the original hens, and as they were still laying, with a second generation cockerel back to the two hens yet again. This process was intended to revive Mr Wright's strain. Then a Sicilian Buttercup hen was included in the strain that brought in willow green shanks and feet, and white lobes. It is not clear exactly when this private flock was launched as a 'new breed'.

A handful of people started to breed them, some close to home in Lancashire, and others down in Somerset. They were first exhibited in December 1920 at Wellington Show, Somerset; and then a few weeks later at the Darwen

(Lancashire) Fanciers Members' Show, January 1921. At the latter event, four keen breeders decided to form the Marsh Daisy Club. This was four out of about twenty known breeders at the time. By December 1921 the club had forty members, and there were about 200 flocks of Marsh Daisies known in the UK. A breed standard was published that was recognized by the PCGB in December 1922.

Mr G.W. Stott of Darwen, Lancashire was the first club secretary, until about 1927/28 when he resigned owing to business pressures – this was the time of the Great Depression. It is not known what, if any, part Mr Moore played in the running of the club for the breed he had made. There were three more secretaries in the club's history: Francis J. Hemelryk (Old Mill House, Partridge Green, Sussex) 1927/28–1930, James P. White (South Road, Watchett, Somerset) 1930–33/34 and D.J. Insley ('Ulva', York Avenue, Hunstanton, Norfolk) from 1933/34 until the club died out during the war. As today's stock was from Somerset, two more leading breeders should be mentioned: Miss Margaret Cavill (Winpenny Farm, near Taunton) and Mr E. Parsons (Fore Street, Williton).

The club started full of optimism, publishing newsletters in 1921, 1922 (500 copies!) and possibly more. There were several successful Club Shows (detailed *below*), but classes for Marsh Daisies were seldom provided at normal local shows, apart from the few areas where there were several enthusiasts in one place:

- 1921, Crystal Palace, three classes, forty-one entries (then three colours, Buff, Wheaten, White)
- 1925, Birmingham, four classes, fifty-eight entries, believed to be their highest ever show entry. (Included Blacks and Browns in AOC, M/F class)
- 1926, Crystal Palace, ? classes, fifty entries
- 1931, Crystal Palace, two? classes, twenty-three entries
- 1932, Crystal Palace, two classes, twenty-eight entries (classes AC, M, AC, F)
- 1933, Crystal Palace, two classes, ? entries. (classes Wheaten, M/F, AOC, M/F).

Even the enthusiasts of the Marsh Daisy Club in the 1930s admitted that their favoured breed was neither a top exhibition breed nor used by large-scale commercial producers, but they laid moderately well for up to five seasons and the cockerels were meaty, so were ideal for smallholders, which is still true today. Marsh Daisies have won mixed Rare Breed classes since the 1980s, but usually at smaller shows.

Buffs, Wheatens and Whites, the three original colours were all made by Mr Moore. There was not a great difference between some Buffs and some Wheatens. Both varieties tended to merge together as Black-Tailed Buffs. Whites were always much rarer. Marsh Daisies are required to have 'willow green' coloured shanks and feet, but most Whites had yellow feet. In *Poultry World* magazine (2 December 1921) Mr Moore advertised Whites for sale, which were bred from: 'The only pen of Whites with willow-green legs in existence.'

Browns first appeared at the 1925 Royal Lancashire Show, where two 'well filled classes' were won by Lord (later Viscount)

Marsh Daisy, large male. Photo: John Tarren

Canterbury, from Norfolk (not Kent, as his title suggests). Blacks were first exhibited about the same time, the leading breeder being Mr E.C. Parsons from Williton, Somerset. One of his Black cockerels was Best of Breed at the 1931 Club Show. It is not known how they were bred. Blues and Exchequers were said to have been 'on the way' but never appeared. There has never been a bantam version.

Description

The standard calls for, 'Carriage: Upright, bold and active', 'Legs: Moderately long' and 'Plumage: Semi-hard, of fine texture; profuse feathering to be deprecated.' Photos of winners back in the 1920s and 1930s clearly show birds that fit these descriptions. Most of today's Marsh Daisies seem rather shorter and more feathery than those, although fanciers should remember that reachy, active-looking cockerels and pullets often look like different birds when they are three or four years old. Chickens get middle-aged spread too. Breeders should study the old photos, and produce enough each season to restore the original type.

Weights range from 2kg (4½lb) pullets up to 3kg (6½lb) adult cocks, and their bodies should appear 'square and blocky'. When seen from above, the body should be the same width along the entire length, and likewise, the back line should be parallel to the keel. When standing normally, the back line should be slightly above the horizontal.

Marsh Daisies have rose combs that should ideally have a fairly short leader, a little above the skull, but not high. Eyes are red and they have white almond-shaped lobes. Shanks and feet should be light 'willow green' on all varieties. Blacks usually had much darker shanks and feet, most Whites had yellow feet. Both colour varieties have been extinct since (±) 1939, so this is purely academic, unless someone is planning a remake.

The three main colour varieties are Buff, Wheaten and Brown, although it is sometimes difficult to tell which of these an exhibit is supposed to be. Buff and Wheaten males are both variations of Black-Tailed Buffs, the former being a lighter shade, buff to the skin and with tails that are bronze, but not black. Wheaten males are darker ('golden-brown'), have grey under-colour and glossy green-black tail feathers. Brown males are similar to Welsummer colour.

Buff females should be buff to the skin, usually with some black tail markings. Wheaten females are similar to some old strains of Oxford type OEG, creamy 'white-wheat' breast and under-parts, darker 'red-wheat' back and wings. They have grey under-fluff, black tips on neck hackles and black tails. The few Brown females shown in the 1920s and 1930s were a variable lot, some similar to Partridge OEG, others with Asiatic pencilling.

NORSKE JAERHØN

Very little information in English is available on this typical small European village fowl. Similar chickens may have been scratching around in Norwegian farmyards for centuries, but they were not made into a standard breed until 1920. They were associated with Stavanger, a town in south-western Norway.

They are compact little birds, with weights from 1.5kg (3¼lb) pullets to 2kg (4½lb) adult cocks. Upright single combs and white ear lobes are required. There are now two colour varieties, both of which are Autosexing; sex-linked barring combined with gold and silver versions of the pattern seen (with differences of detail) on Friesians, Pencilled Hamburghs, and so on, Silver Norske Jaerhøn males are mostly white with black (or grey) and white cuckoo barred tail and wing markings. Gold males are similar with some yellow/gold feathering, mostly on the saddle. The females are much darker, black (or grey) on white or black (or grey) on gold barring on all parts except for the clear silver or gold head. The author does not know how these presumably amateur breeders in relatively remote Stavanger, produced an Autosexing breed almost a decade before some of the world's leading geneticists at Cambridge did so.

OLD ENGLISH PHEASANT FOWL

This breed officially came into being in 1914, but was really a revival of the original Gold Spangled Yorkshire Pheasant and Lancashire Mooney Fowls that had existed in the early nineteenth century. Most of them had been absorbed into Gold Spangled Hamburghs. There had been competitions for various local breeds in the counties of northern England before the start of general poultry shows in 1845. These larger events increased the level of competition, and the size and shape of the spangles and other markings had to be perfect (according to the revised standards) to win. Exhibition Gold and Silver Hamburgh hens now need large circular spangles, unlike the half round, or 'Moony' spangles of the old strains. Mr Beldon (quoted in *Wright's Book of Poultry*) said that by 1873, 'In Yorkshire also, though I live there, I should have some difficulty now in finding a pure-bred Pheasant.' However, a few did survive on remote farms in the Yorkshire Dales. A new generation of historically minded breeders revived the old strains in the early years of the twentieth century, and in 1914 the 'Pheasant Fowl Club' (later 'Old English Pheasant Fowl Club') was formed. There had been some argument about the name, some had preferred 'Yorkshire Pheasant Fowl Club'.

With the general disruption to everyones' normal lives during the First World War, plus whatever personal circumstances caused the first club secretary, J.S. Cayley, to move house several times in a few years, the club existed in name only by 1920. Several breeders met in 1922 to revive the club, with George Dent of Wigston, near Leicester becoming the acting secretary until Mrs Synge of Chester held the reins from 1923 to 1933. Mr J.W. Parkin, of Mickleton, Yorkshire, then became secretary until the club finally closed down about 1960. OEPF have been covered by the Rare Poultry Society since it was formed in 1969.

Between the wars, Pheasant Fowl remained a dual-purpose smallholders' breed in their original home region and other exposed places around the country. They were not used by larger-scale commercial egg producers, despite giving some good performances at a few of the laying trials, one hen ('SP 96') turning out 253 eggs in forty-eight weeks. Surplus cockerels were not as well-proportioned as the specialist table breeds, but were good enough for the local markets.

Although OEPF breeders wished to emphasize the hardiness and productivity of their birds, the club members did have an annual show most years, usually at Birmingham, and supported a few northern shows in good numbers. In 1923, there were forty-eight at Egremont Show and fifty-eight at Ulverston Show. The two highest entry figures for the main Birmingham event found were fifty-one in 1933 (ten cocks, eleven hens, fifteen cockerels, fifteen pullets) and fifty-nine in 1934 (twelve cocks, ten hens, twelve cockerels, fifteen pullets, ten selling classes). Harold Easom Smith recalled (in *Modern Poultry Development*, 1976) that the last big entry of them he judged was at the 1956 OEPF Club Show at Leeds. Since then there have seldom been more than a dozen at any one show; and the total population has remained alarmingly low.

There has never been a bantam version of this breed, and as a traditional utility breed, there would seem no point in making any. Bantam breeders already have Gold Spangled Hamburgh and Derbyshire Redcap Bantams, and there are not many of either of these.

Description

In order to maintain egg size and table qualities, size is important. Pheasant Fowls range from 2kg (4½lb) pullets up to an impressive 3.2kg (7lb) for adult cocks. Compare this with the much smaller Hamburgh cocks at 2.24kg (5lb) and slightly smaller Redcap cocks

Gold Old English Pheasant Fowl, large male. Photo: John Tarren

Gold Old English Pheasant Fowl, large female. Photo: John Tarren

at 3kg (6½lb). Their body is rather long, with a meaty breast and prominent shoulders. Cocks have large fanned tails that are described in the standards as 'set well back', but are actually often carried quite high. Hens do usually have lower tails that are 'moderately whipped'.

Pheasant Fowls' rose combs are normally larger than those of Hamburghs, but naturally considerably smaller than Derbyshire Redcaps' combs. The leaders (rear spike) of OEPF's combs are quite long and should follow the line of the skull, but not touch it. They have red eyes and white, oval ear lobes. This provides another obvious distinction from Hamburghs that have larger, circular lobes and Derbyshire Redcaps that have red lobes.

A Silver variety is recognized in *British Poultry Standards*, but few have been seen since the breed was officially revived in 1914. Mention Old English Pheasant Fowl to most poultry people, and they will automatically think of Golds. The ground colour is composed of several shades of 'bright rich bay', a reddish-brown colour. Males have glossy green-black tails, hackle striping and breast markings. Their breast feathers are completely laced (as Wyandottes), a common cause of confusion among general judges who expect them to have the same crescent-shaped markings as females.

They often diverge into two opposite, equally undesirable, directions, either too light or too dark. Those at the light end are often good-sized birds with the correct markings, but fail on ground colour. The dark birds, for some reason often under-sized, have the correct rich ground colour, but have too much black. Spangling and lacings are too heavy. It should be noted that although Pheasant Fowl hens have very similar markings to Redcap hens, the males are noticeably different, Pheasant Fowls have laced breasts and Redcap males have solid black breasts.

Special Management Requirements

To keep them fit, they need extensive free-range conditions. These active birds will not do well inside, or if the run is a small area of mud. Unless you are very methodical and well organized, it is probably not a good idea to keep both Pheasant Fowls and Derbyshire Redcaps.

Silver Old English Pheasant Fowl, large male. Photo: John Tarren

OSTFRIESISCHE MÖWEN/ EAST FRIESIAN MÖWEN (Möwen = Gulls)

Ostfriesland is the northern coastal district of Germany just over the border from the Dutch Groningen district. Because both areas face the North Sea, it must be assumed that they would need to be hardy to survive; their coastal homeland must be the source of the sea-bird inspired names. German Ostfriesische Möwen and Dutch Groninger Meeuwe are very closely related, and both are also related to Dutch Fries Hoenders (Friesians).

These three (now) recognized breeds would have all been one variable population before the age of shows and standards. Going further south, and allowing for rose combs as well as singles in mixed farmyard flocks, a host of other breeds, including Assendelftsers, Brakels, Campines and Hamburghs can be added to the relations; and similar birds would have existed in medieval times. However, Edward Brown (in *Races of Domestic Poultry*, 1905) could not give any more specific history beyond stating that the breed's name had only been used since the 1880s. Sometime later, possibly circa 1914, Ostfriesische Möwen (single-combed) were differentiated from the generally similar, but rose-combed, Westfälischer Totleger.

Ostfriesische Möwen are kept by a limited number of enthusiasts, mostly in their home region. Rex Woods (in *Rare Poultry of Ashe*, 1976) said that he had seen them in South Africa, but apart from these, there are unlikely to be many kept anywhere else in the world. At the time of writing (2005), none were known in the UK.

Ostfriesische Zwerg-Möwen/East Friesian Möwen Bantam

These are a recent creation in comparison to other breeds of poultry. Silver Ostfriesische Zwerg-Möwen were recognized in 1960, with the Golds following in 1968. L. Groen from Elisabethfehn and E. Flemer of Aurich made and perfected Silvers, initially from crosses of large Silvers, Silver Partridge Zwerg-Italiener (equivalent to UK Silver Duckwing Leghorn Bantams) and (Silver Duckwing?) Deutsche Zwerghuhner (German Bantams).

Although not standardized until 1968, Golds were first exhibited in the new/non-standard section of the 1964 Osnabröck Show. Breeders mentioned as being involved in making Golds, include Mr Onken from Engels, Ostfriesland, Dr Hildebrand from Köln and E. Flemer as above. The then new Silver Ostfriesische Zwerg-Möwen were crossed with appropriate other Gold bantams, including Gold Friesian and Millefleur Rosecomb Bantams.

Description

This is a light breed, but a very compact, plump-bodied one. Large fowl range from 1.75kg (3¾lb) pullets up to 3kg (6½lb) adult cocks. Bantams are 700g to 900g (25–32oz). These weights are all higher than those of its Dutch relation, the Groninger Meeuwen.

Combs are single, straight and upright. Ear lobes are white and oval. Eye colour is orange to reddish-brown, a distinct difference

European Light Breeds

Gold Ostfriesische Möwe, pair. Artist: Kurt Zander

from the dark brown eyes of Groninger Meeuwen. Shanks and feet are slate blue.

For most of their history, there were just two colour varieties, Gold and Silver Pencilled. In more recent years, Gold/Blue Pencilled, Silver/Blue Pencilled and Buff/White Pencilled have been added to the range. No doubt if it proved possible to make any more colour combinations, these would be accepted, but only in the one pattern. This pattern is a coarse variation of cock-breeder Pencilled Hamburghs, Pencilled Friesians or Sicilian Buttercups.

Taking Silvers as an example, most of the plumage of males is silvery-white, except for the tail, which is mainly black. There are a few black markings on the wings and the thighs. Main tail feathers have white shafts and just a little white edging, but are never as well defined as on the best Exhibition Pencilled Hamburgh males. Some edged saddle hackle feathers are quite common however.

Females have the 'ears of wheat' design seen on Buttercups and Friesians. The neck and upper breast is clear, with fewer markings on the lower breast and belly than on the back and wings. Tail feathers should ideally have the same 'ears of wheat' design, but as with many cock-breeder Pencilled Hamburgh hens, they sometimes have markings along the length of the feathers instead.

PENEDESENCA

This breed from the Catalan region of Spain breaks at least two of the 'normal rules' of the poultry-keeping world. Brown-shelled eggs are normally associated with red ear lobes, but Penedesencas have white lobes and lay very dark brown eggs, usually darker than those of Marans or Welsummers. Secondly, side sprigs on single combs are regarded as a

Silver Ostfriesische Möwe, large male. Artist: Kurt Zander

Silver Ostfriesische Möwe, large female. Artist: Kurt Zander

disqualifying fault on every other breed, except for these and their Catalan relation, the Empordanesa. This was written before a standard for Penedesencas had been presented for inclusion in British Poultry Standards by the PCGB Council. The likely reaction of some Council members, on learning that large side sprigs are compulsory on these two breeds, will probably be extreme.

No details of their early history were found by the author, but the best guess is that they arose from crosses between typical white-lobed, white-egg laying, Mediterranean type chickens and some now unknown imported Asiatics in the early twentieth century. Their very existence, and possible economic potential, were stopped by the disruption of the Spanish Civil War and then the Second World War.

The revival of the breed from (presumably) a few survivors found on farms in Catalonia started in 1982 at the poultry section of the Insitut de Recerca i Tecnologia Agroalimentàries in the region, Amadeu Francesch and Antonio Jordà being the leading scientists of the project. Since they have made stock available to the public, most of the contacts and information have been via the internet, which explains why this technophobic author hasn't found much information on them. Live specimens have followed the flow of information, and small numbers of them are in the USA (and Canada?) and much of the EU, including the UK.

Description

They are a typical light breed in general size and shape, with weights ranging from 2kg (4½lb) pullets up to 3kg (6½lb) adult cocks.

European Light Breeds

The controversial comb shape can either be a mass of side-sprigs ('carnation comb') or form a Christian cross formation when viewed from above, with one large sprig on each side.

There are four recognized plumage colour varieties, Black, Partridge, Wheaten and Crele. Blacks and Wheatens do not need further description. Partridge males are normal Black-Reds, and the hens are Asiatic pencilled (as Partridge Cochins and so on). Creles are as one would expect from the combination of barring and this type of partridge pattern. Shanks and feet are nearly black on Blacks, blue-slate on Partridge and Wheatens and white on Creles. When selecting future breeding stock, pay at least as much attention to eggshell colour as plumage colour and so on. Be aware that there may be a trade-off between egg numbers and eggshell colour; there is a limit to the amount of pigment a hen can make.

RAMELSLOHER

Although Germany is now the centre of hobby poultry keeping (the shows there are far bigger and better organized than anywhere else in the world), it was not always so. Poultry showing, and hence the standardization of breeds, started about 1850 in Britain and America. Poultry keeping was quite backward in Germany then, and it was not until the 1870s that they started to catch up by standardising existing local breeds and making new ones. Ramelslohers, along with Rheinländers and Sachsenhuhners that follow, are all German breeds that were made in the late nineteenth and early twentieth centuries. None of these three breeds made much commercial impact in Germany, although Rheinländers eventually attracted a lot of fanciers. Many German poultry keepers, both farmers and hobbyists, took up 'international' breeds with great enthusiasm, especially Wyandottes.

The Ramelsloher was developed and made by N.D. Wichmann, presumably as a hobby since he was described as a shipowner from Hamburg. He bred them for some fifty years, roughly 1875 until 1925. It appears that Herr Wichmann's idea was to refine and stablize utility cross-breds that many others in the region kept as well.

The city of Hamburg was growing, like major industrial cities and ports everywhere in Europe, and with it was an increasing demand for food for the people, including poultry-meat and eggs. To improve productivity, local breeds (in this case Silver Ostfriesische Möwen) were crossed with imported Minorcas. Ramelslohers were first exhibited in the new breeds section of the 1874 Hamburg Show. Whites were to become the most popular colour by far, with Blacks and Speckleds making a brief early appearance. Buffs, usually of a light shade, were another early variety (circa 1875) that still exist today, and were first made from crosses between White Ramelslohers and Buff Cochins. Blues were a short-lived variety, made about 1890 from Andalusian crosses.

During their period as a commercial breed, roughly 1880 to 1914, they had a reputation for being early layers and free breeders, producing eggs and table cockerels in the late winter and early spring when prices were highest. The author has not found any details of the housing used, but would guess that the birds must have been kept inside. However, the feeding method has been recorded, warm winter meals of ground buckwheat, fish meal (from processing plants in Hamburg and along the Elbe?) and curdled milk.

No details have been found, but the author would guess that Ramelslohers had a chequered history from 1914 until 1945; although being a distinctly German breed, the Nazis probably encouraged their breeding from 1933 until they started losing the war, about 1943. In modern times, Ramelslohers have not attracted a lot of fanciers. Some major shows during the 1970s and 1980s, did not have any entered. They have survived through to the present, and as long as interest in pure breeds of poultry continues at

White Ramelsloher, large male. Artist: August Rixen

White Ramelsloher, large female. Artist: August Rixen

something like today's level, should continue to do so. None are known in the UK, and although they are attractive and stylish birds, they do not seem to have any unique features likely to tempt many British breeders to swap from whichever breeds they already have. It would certainly be interesting to us 'Rare Breed Anoraks' to see some at UK shows, but any that do appear will probably belong to fanciers with some German connections.

No references to a miniature version of Ramelslohers have been found.

Description
Of the breeds used to make them, it is the Minorcas that are most obvious. They have the same height and impression of activity, differing in having much smaller lobes, combs and wattles. Plumage is fairly tight, which adds to their athletic appearance. Neck hackles do spread far over the shoulders or back, which combined with close body feathering and a tail that is only medium in length, moderately fanned and held 'well out', makes the back look very long. Body weights range from 2kg (4½lb) pullets up to 3kg (6½lb) adult cocks.

Only the White and Buff varieties are seen now, the Buffs being of a fairly light shade. Both varieties have bluish-grey beak, shanks and feet, medium/small bluish-white ear lobes and very dark eyes, usually associated with dark pigmentation of the skin around the eyes.

RHEINLÄNDER

These are from the part of Germany over the border from the French region of Alsace, so the German Rheinländer and French L'Alsacienne chicken breeds are very similar. It is not known if any birds have been exhibited as both breeds, perhaps in a French show one week and a German show the next.

European Light Breeds

Rheinländers are a relatively new breed, the originator Dr Hans Rudolf von Langen from Köln started to make their forerunners in 1894. His first birds were of the Silver Partridge (aka Silver Duckwing) colour variety, and called 'Silberhalsige Deutsche Landhühner'. Dr von Langen made them by crossing Rebhuhnfarbige Italiener (Brown Leghorns), Ramelslohers, Bergische Krähers, French Le Mans and a local breed, believed now extinct, from the Eifel district of Germany near the border with Luxembourg, halfway between Köln and French Alsace. His idea was to combine the productivity of the Italieners/Leghorns and Ramelslohers with the hardiness of the local breeds. The first birds were in the Silver Duckwing plumage colour variety, and were shown as 'Silberhalsige Deutsche Landhühner' in the 1897 and 1898 German National Shows. They were renamed as 'Rheinländer' in 1908.

Although the 'Silberhalsige' variety came first, they have become very rare since then. Blacks soon became established as the main variety, this colour no doubt coming from the Le Mans, as did the rose comb genes for the breed. White Rheinländers probably came from Ramelslohers, Blues from Andalusians and Barreds from Deutsche Sperbers.

Huge entries of Black Rheinländers can be seen at the main German shows, often well over 100 large Blacks. Typically there will be about ten of each of the other large fowl colour varieties. There are a fair number of Rheinländer breeders in The Netherlands, but apart from these they do not seem to have become very popular around the world. A few were seen at UK shows during the 1990s, when someone brought over some eggs, but they seem to have died out.

Black Rheinländer, large pair. Artist: Kurt Zander

Zwerg-Rheinländer/ Rheinländer Bantam

Ewald Camphausen of Bamenohl, Sauerland, is credited as the originator of the miniature version of the breed, starting in 1918. He used large Black Rheinländers and various black bantams, including Deutsche Zwerghühner, OEG and crossbreds. They were first exhibited at Leipzig in 1921, but were not officially standardized until 1932. The White variety was added in 1938. Progress was obviously halted by the Second World War, and so had to wait until about 1950. By 1963 their popularity had risen dramatically, and was illustrated by a display of 320 of them at Hildesheim Show.

Description

When seen in profile, the body should have the appearance of a round-cornered rectangle in the ratio of 8:5 (length : depth). Both sexes have large, well fanned tails, carried at about 45 degrees above the back-line, and the back is carried just above the horizontal. They are still a light breed, but their depth of body make them look almost like Dorkings.

Their headpoints are very distinctive, combining smallish, smooth-topped rose combs (as otherwise usually seen on some German strains of Wyandotte Bantams) combined with modest sized, broadly oval, pure white ear lobes. Wattles are rounded (not long and pendulous) on males, small on females. The large fowl range from 1.75kg (3¾lb) pullets to 2.75kg (6lb) adult cocks; with bantam males at 900g (32oz) and females 800g (28oz).

Plumage colour varieties are, in approximate order of popularity: Black, White, Laced-Blue (as Andalusians), Black-Red/Partridge, Silver Partridge (Silver Duckwing) and Barred. Eye colour is as dark as possible. Shanks and feet are black on young Blacks, fading to slate-blue on older birds, which is also the standard colour for all the other varieties except for Barreds that have white shanks/feet. It is impossible to obtain slate-blue shanks/feet on any chickens with barred or crele plumage.

SACHSENHÜHNER

(These could be called 'Saxony Fowls' in English, but the German name is usually used.)

Starting in 1886, a series of matings involving Minorcas, Croad Langshans, Plymouth Rocks and Sumatras were started, with the intention of making a breed that would retain the egg production abilities of the Minorcas, but with more compact, better table quality cockerels. This was in the Erzebirge area of Sachsen (Saxony) region, between the cities of Chemnitz, Dresden and the present (2005) border with the Czech Republic. Other breeders were working on similar lines in parts of Bayern (Bavaria). These areas have cold winters, so another aim was to have smaller, less frostbite-prone, combs, lobes and wattles than Minorcas. Sachsenhühner were first exhibited at Dresden in 1905 and Nürnberg in 1908.

A breed club, the Landesverbond Sächsische Geflügelzuchter Vereine, was formed in 1914, and a breed standard was published. As might be expected, the First World War slowed progress, but people still kept chickens, and Sachsen/Saxony was a long way away from the Western Front. By 1923, Arthur Esche of Chemnitz-Ebersdorf had made Barred and White Sachsenhähner from crosses between the original Blacks, Plymouth Rocks (Barred? White?), Dominiques, Deutsches Reichshühnern, White Rheinländers and White Leghorns. The Buff variety did not appear until 1960 and was made from crosses between existing (White?) Sachsenhühner, Buff Orpingtons and Buff Italiener (German equivalent of Buff Leghorns). They are rare in Germany, and virtually unknown elsewhere.

Zwerg-Sachsenhuhn/Saxony Bantam

These are a recent creation, first made by Franz Bilzer of Trebur in the early 1970s. Large Black Sachsenhühner were crossed with Black Rheinlander Bantams, Australorp Bantams, and Deutscher Zwerghühner

European Light Breeds

Black, Barred and White Sachsenhühner, pairs. Artist: Kurt Meißner

(German Bantams). He made a second strain during the 1980s, again using large Black Sachsenhühner and Rheinlander Bantams, but this time the other bantam breeds used were German Langshan Bantams, Black Italiener/Leghorn Bantams and Augsburgers (large? bantam?). Barred and White Zwerg-Sachsenhühner also exist, but the author does not know if there is a miniature version of the large Buff variety.

Description

Sachsenhühner have small white lobes, upright single combs, fairly small wattles, medium to shortish legs, and fairly long tails carried about 30 degrees above the horizontal.

The main colours are Black, White and Barred. Buffs are not quite self Buff, neither are they quite Buff Columbians either, but are somewhere in between. They have some bronzy shading in tails, neck and saddle hackles. Shanks and feet are black on Blacks, white on all other varieties.

Standard weights range from 2kg (4½lb) pullets up to 3kg (6½lb) adult cocks for the large fowl, with ideal miniatures being 900g (32oz) for the males and 800g (28oz) the females.

SCOTS GREY

Although regularly described as the traditional old breed of Scotland, there is not very much detailed documentation about them before the 'Scotch Grey Club' (changed to 'Scots Grey' circa 1923) was formed on 9 December 1885 at The Waverly Market Christmas Show, Edinburgh. The breed certainly existed long before then, possibly as far back as the eighteenth century, but the few descriptions found suggest they were once heavier birds with fuzzier cuckoo markings than the sharply barred, elegant fowls favoured by fanciers since 1885. There was a gradual change, and some strains may have had a continuous history over the period of change. Professor Stephen Young of Glasgow University was a well-known enthusiast who mentioned to Harold Easom Smith (ref: *Modern Poultry Development*, 1976) in 1950, that he had a medal that had been won by a family with Scotch Greys in 1864. As this was before the breed club was formed, the medal may have been won at a Scottish agricultural show.

James Long (*Poultry for Prizes and Profit*, 1886) said hens weighed 6lb (2.7kg) to 7lb (3kg) and cocks 8lb (3½kg) to 9lb (4kg). He continued, 'The tail is flowing, and the general outline of the bird is somewhat similar to that of the white Dorking.' Edward Brown made similar comments in his 1893 book, *Pleasurable Poultry Keeping*, that were very different from the descriptions in his later books.

Harrison Weir, in his *The Poultry Book* (UK 1902, US 1905), mentioned seeing some good quality, uniform flocks of them about 1860. He appears to have noticed two distinct types, 'the old sort – square, plump, short in leg and thigh, with medium length of shank', and remarked that 'the true old Scotch Grey was a trifle more Gamey and upright in its carriage'. It is probably the case that the latter type became the current Scots Grey and the plumper type were ancestors of Scots Dumpies, but without the extreme short-legged Creeper gene.

The majority of Scottish poultry keepers were cottagers, and they were only interested in egg and meat production, plus broodiness and mothering ability. Dealers did their rounds of the crofters, taking produce to the cities, Glasgow, Edinburgh and Berwick, where some was sent by train to London.

It seems odd today to think that it would have been viable to transport eggs all the way from remote Scottish crofts down to London, especially considering the number of specialist producers there were in the much warmer counties of Kent, Sussex and Surrey, plus the not too distant duck producers around Aylesbury and the considerably shorter train ride away to farmers in Lincolnshire and East Anglia. In fact it was the poverty and poor housing of the Scottish cottagers that gave them an advantage during the cold weather of winter.

In medieval Britain, from the Scottish Highlands to Kent, almost everyone would have lived with their livestock under the same roof. Chickens would have roosted on roof beams, and so enjoyed the heat and light from the fire. Most families only kept a few birds, so it would not have been too bad. The birds would also have been given warm mashes of oatmeal, or whatever was available. This way of life was still quite common in Scotland and remote parts of England and Wales in the nineteenth century, long after it was abandoned in the more prosperous English counties nearer London. Chickens in these counties would have all been shivering in hen coops, and not laying much as a result, during the winter. Therefore, the price of eggs in London went high enough for it to be worth transporting them down from Scotland.

The majority of Scots Grey breeders have always been in Scotland, although there has consistently been a number of English supporters, often those living in bleak and exposed places. As far as is known, very few have been exported around the world. The 1922 *Feathered World Yearbook* mentioned an expat in the Indian Himalayan mountains who found them the only breed to survive up there.

Large Scots Greys (*see below* for the bantam version) have been bred in limited numbers, even in Scotland, from 1945 until the present day. This leaves a clearly defined period of relative popularity for the breed – 1885 to 1940. For over half of this period, from about 1910 until his death in 1940, John Carswell of Falkirk, was Club Secretary. A main Club Show was held almost every year (1914–18?) in Scotland. On many occasions, it

Scots Greys, pair. Artist: J.W. Ludlow. This picture was originally a free gift with Poultry *magazine, circa 1912–14.*

was on 1/2 January at Kilmamock. We would imagine that the birds were penned the day before, and all the exhibitors celebrated Hogmanay together at one or more local hotels.

It has not been possible to find complete records, but their most successful show was probably the January 1912 event, when for the first time an Englishman, F.J.S. Chatterton was invited to travel up to Kilmamock. As he lived in Finchley, London, it may have been a long way, but it was a simple train journey. Awaiting him were twelve cocks, nine hens, twelve cockerels, seventeen pullets in the main competitive classes and an additional twenty-two in two selling bird classes. The Scotch Grey Club did not offer classes for their bantam version until the 1920s. Everyone was said to have been very pleased with Mr Chatterton's judging and the 'birds were big and shapely, as well as being the acme of perfection in colour and markings'. Alexander Hamilton, then the oldest leading breeder, said this was the best all-round show to date; meaning slightly greater numbers may have appeared before, but of poorer average quality.

Between the wars, the best Scots Grey Club Show was probably their December 1931 event at Glasgow. Here there were sixty-three large, plus now fourteen bantams. No details of the bantam classes have been found, but the large fowl comprised: seven cocks, fourteen hens, sixteen cockerels, eighteen pullets and eight selling. The last of this series of shows was at Lanark in 1939. The club folded during the Second World War, not helped by the death of John Carswell in 1940. It was not re-formed until 1970, when Arthur Frame from Biggar became secretary, remaining so until his death in 1978.

Since 1970 more members have been keeping bantam Scots Greys than large ones. By 2002, the DEFRA (UK government agriculture department) census of livestock breeds estimated (no exact figures exist for poultry breeds) that there were only about 100 breeding females in the country. This included those kept by the Scots Grey Club members, farm parks, and a few others. There may have been more in back gardens somewhere, but there would not have been enough others to make much difference to their rarity. At the 2004 National Show at Stoneleigh, there were twenty (eleven female, nine male) large entered, and this was more than had appeared in the previous few years.

Scots Grey Bantam

The date of origin of large Scots Greys may be completely unknown, but that of the bantam version is not quite, but nearly, known. Bantams with cuckoo barred plumage have obviously existed for a long time. The ancestors of Cuckoo Belgian Barbu d'Anvers were depicted in a painting in 1817, and may have been one of the source breeds used by the first recorded Scottish exhibitor, Mr Mitchell of Paisley. He kept them at least as far back as 1866. Apparently several other breeders had the same idea around the same time, including Dr. Boulton of Beverley (Yorkshire, England), Henry Beldon, Mr Phelps (Ross, Dumfries and Galloway, Scotland), Miss Jackson and Mr Edwin Wright.

Scottish breeders intended to produce a miniature version of the large Scots Grey from the beginning, but many English authors simply referred to 'Cuckoo Bantams'. Some had rose combs and white lobes, and might eventually have become Cuckoo Rosecombs here in the UK if the British Rosecomb Club had been more open to new varieties. Others were of the same body shape, but with single combs, and no doubt were identical to today's Cuckoo Dutch Bantams. Still others were thicker set, with more profuse plumage, similar to Scots Dumpy Bantams.

They remained, as far as show entries indicate, in very few hands until the 1950s. There may be more now than ever before, although there may have been a lot of unrecorded flocks in earlier times. At the 1921 Crystal Palace Show there were six males and six females, the highest show entry found until the Scots Grey Club Show, Christmas 1932 (Glasgow?), where nine males and nine females appeared. In earlier times, the (then) 'Scotch Grey Club' did not provide classes for the bantams, although several members kept them. In the 1 February 1901 issue of *Poultry* magazine

Scots Grey, large male. Photo: John Tarren

Scots Grey, large female. Photo: John Tarren

European Light Breeds

Scots Grey, bantam male. Photo: John Tarran

Scots Grey, bantam female. Photo: John Tarran

there was a letter from F.J.S. Chatterton, then living at 78 Clissold Road, Stoke Newington, London (before he moved to Finchley) that announced the formation of a Scots Grey Bantam Club on New Year's Day at Paisley Show. He was to be the Secretary, Matthew Smith the President, J.R. Robins the Vice-President, plus a Committee of three, James Braidwood, Thomas Lambie and Alex Meikle. Despite the enthusiasm of the founders, for some reason this bantam club did not last, and was certainly defunct by 1910. Compare this low level of interest in Scots Grey Bantams with the PCGB National Show at Stoneleigh Warwickshire (a long way from Scotland) in 2003, where thirty-eight were entered (eight cocks, nine hens, eleven cockerels, ten pullets).

Description

Close study of paintings and drawings that usually showed someone's ideal, and photographs that show the reality, reveal considerable variation in the shape and size of Scots Greys over the last century or so. According to descriptions, they were once much heavier than they became later as an exhibition breed. Engravings by Ludlow and others, show the markings were much fuzzier than would have been considered acceptable after about 1900. By the 1930s, some had gone in an almost opposite direction – that is, with wonderfully clear barring, but on birds that were elegant to the point of being scrawny. Large males should weigh 3.2kg (7lb), females 2.25kg (5lb), and it is still the case that the largest birds usually have poor barring. Scots Grey Bantam males should weigh 620–680g (22–24oz), females 510–570g (18–20oz). Many are too large.

They have fairly long bodies that, combined with their fairly tight plumage, gives a long-backed appearance. The breast should be full and rounded, and the back reasonably broad. Wings should be carried 'well up', which is most likely to be a problem on the smallest of the bantams. Oversized bantams usually have correct wing carriage.

Head points are simple, medium-sized straight single combs (allowed to fall over slightly on females) with medium-sized red lobes and wattles. Eye colour should be amber, and the beak should be white, frequently with some black streaks. Similarly, their shanks and feet are either white or white with black spots.

Barring on the plumage should be as clear and sharp as possible. The black bars should have a glossy 'metallic lustre', and to obtain this the white bars are allowed to be pale grey. The bars should be of equal width, or if not, the black should be slightly broader than the (nearly) white. This is most likely to be a problem on the neck and saddle feathering of males that are often too light in general shade. Whilst most of the barring is straight across the feathers, male neck, saddle and tail feathers usually have V-shaped bars. One fault of which fanciers should be aware, is that wing flight feathers with barring do not go right across the feather, but instead fade out to a fuzzy grey at the edge.

SPANISH

This is the author's favourite breed, having bred large Spanish since 1978 and the bantam version since 1988, so it has been very frustrating not to have found full details of their origin. Despite their name, it is most unlikely that they were developed in Spain. The 'raw materials' would have come from Spain, but they were almost certainly red-faced, white ear lobed, black-plumaged fowls similar to Castilians. During the centuries when White Faced Spanish might have been first developed, say from 1580 to 1780, poultry keepers in Spain kept either layers or fighting cocks, and only the latter were carefully bred by their owners. The layers had a good reputation for high production of large eggs, so many were exported to northern Europe. From about 1500 to 1700 the Hapsburg dynasty ruled Spain and (present-day) Belgium and The Netherlands, obviously increasing general trade between these countries. Religious history is also important in this context; the Reformation gave Protestant breeders the freedom to openly begin selective breeding of plants and animals.

Under Roman (theological) rule, anyone rash enough to announce that they had created a new variety of auricula, chicken, canary, rabbit, tulip or whatever would have been accused of heresy, tortured, tried, found guilty and then executed in a painful and publicly entertaining way. Everyone expected the Spanish Inquisition! The division between the Catholic and Protestant Churches was protracted and violent, and when the Catholic Church attempted to reassert its authority, many were forced into exile, notably the Huguenots who fled to England from 1570.

With all this going on, it is perhaps not surprising that exact details of the origin of a breed of chicken is hard, or even impossible to find. Black Spanish Fowls were noticed, along with the auriculas, tulips and unique hunch-backed Belgian Canaries as being favourites among Flemish refugees (many were weavers) in Bristol and London at least as far back as 1750. The Black Spanish Fowls then are believed to have had white faces, but much smaller than those of show winners 150 years later. White faces may have been favoured because they were thought to be associated with high egg production. Many white-lobed breeds go partly white in the face with age, but Spanish have complete white faces at maturity, about six months of age. It is not known when this was first fixed; it could have been any time after 1600.

The development or origin of the breed was still controversial in the first group of major poultry books published between 1845 and 1875. Charles Darwin's *Origin of Species* and *Variation of Animals and Plants Under Domestication* represented revolutionary ideas that were not accepted by many who still had literal religious beliefs. Some authors, especially 'Martin Doyle' (*nom de plume* for William Hickey, *The Illustrated Book of Domestic Poultry*, 1854) suggested places from Cuba to Cairo as the place where God first put White-Faced Black Spanish. As he even mentioned Noah's Ark, perhaps he thought Cairo was more likely.

It is unfortunate for twenty-first century students of poultry history that Doyle/Hickey and others (from our perspective) wasted so much space on these theological questions when they might have concentrated on what was then recent and easily obtainable history. Earlier editions of *Wright's Book of Poultry* briefly mentioned competitive events held by Spanish breeders long before the first major multi-breed poultry show in the grounds of London Zoo in 1845:

> Many private or "club" shows even then [circa 1800?] took place, being chiefly held in the large rooms of taverns, and mostly confined to some one breed, of which the various amateurs in a neighbourhood thus met to compare their respective achievements, and contend for the pre-eminence in friendly rivalry. By such means the Spanish fowl was raised many years ago to a very high degree of excellence, though differing in several important particulars from the present accepted standards. London was by far the chief home of the Spanish "fancy" in those days.

Later in his Spanish chapter, Lewis Wright went on to say something about the other main centre of the breed:

> The city of Bristol has now for many years [but after they peaked in London?] been known amongst poultry-fanciers for its fine strains of Spanish fowls, a result which very remarkably shows the great influence which may be exerted by a single successful amateur; being almost entirely owing to the sound judgment and perseverance displayed many years ago by Mr. Rake.

Mr Rake retired from Spanish breeding about 1860, so assuming he was elderly by then, he may have started keeping them as early as 1810. His flock was sold, most birds going to leading Bristolian fanciers, Messrs Hyde, Jones, Lane and Parsley.

Spanish were one of the most popular breeds. In the early days of showing, at one of the early Birmingham Shows, circa 1848–50, four classes of Spanish had a total of 103 entries: fifteen cocks, forty-six cockerels, thirteen hens and twenty-nine pullets. In the 1914 *Feathered World Yearbook*, Joseph Pettipher recalled the 1857 Crystal Palace Show, where a total of 240 Spanish appeared over three classes, adult trio, young trio and single cock or cockerel. Shows up to 1860 consisted mainly of trio classes. By the 1870 Crystal Palace Show numbers were down, and the classes had changed, seventy-five birds in six single bird classes (cock, hen, cockerel, pullet, selling male, selling female), rising again in 1885, where 129 entries were spread over the same six classes.

Over the period from (about) 1840 to 1900 the size of their white faces was greatly increased. At the beginning, the faces on typical cockerels reached about half way down their wattles. By 1870 the faces and wattles were about level; and after 1900 photographs of champion males show they had enormous faces. Some were measured, and found to be nine inches (23cm) from the base of the comb down. Winning strains were kept exclusively for showing, and the hens ceased to be a serious commercial laying breed. The skin between the wattles has been required to be white since about 1870; it was red here before 1850.

No details have been found about early Spanish Clubs, either the local Bristol and London clubs, or a national club, assuming there was one, after regular shows became established during the second half of the nineteenth century. Some issues of *Feathered World* from 1892 in the author's collection included results of the Liverpool Show, including seventeen cocks and twenty-five hens, the winning hen, Mr Doble's, won 1st and Special. In another issue, F. Harvey of North Street Lostwithiel, Cornwall, advertised a cockerel that won the 'Challenge Cup' at the 1891 Crystal Palace Show. The existence of 'Specials' and 'Challenge Cups' suggest there must have been an active Spanish Club to award them. F.M. Chatterton was Club Secretary circa 1890–1900, but it is not known when the club was founded.

Feathered World Yearbook did not include the Spanish Club in its lists of breed clubs until the 1921 edition, when J.M. Cooper of 5 Gordon Road, Herne Bay, Kent took on the job of Hon. Secretary. The Spanish Club may have survived in a suspended state up to about 1905. Mr Pettipher mentioned in the 1916 *FWYB* that there a Club Show nearly happened at the 1914 Crystal Palace, but was cancelled because of the outbreak of the First World War. A Mr Whittam was named as Club Secretary, which makes one wonder why the Spanish Club was not listed. If it had gone ahead, some thirty to forty birds were expected, and new cups had been promised by a prominent American fancier, R.A. Rowan of Los Angeles, California. For now unknown reasons, the 1922 *FWYB* gives a new secretary, W. Burton of Burford, Wymondham, Norfolk. He only lasted until 1926. The *Feathered World Yearbooks* (1910–37) collectively provide a detailed account of the British poultry show scene, but Spanish are not mentioned in the 1927 issue, or any others thereafter.

Spanish almost became extinct, but a few survived through to the 1960s in England and Germany. Then interest in old breeds was increasing, and Rex Woods, who had a large collection of breeds at Ashe House, near Axminster, Devon managed to obtain a few inbred Spanish that he crossed with Minorcas to revive vigour and fertility. This cross obviously required several further years of selective breeding to restore completely white faces. At the Rare Poultry Society's first show in 1970 he won the first Spanish class held in the UK for many decades, possibly since 1925 or 1926. Two other UK-based breeders involved in this revival were Fred Tyldesley of Hawkshaw, near Bolton, Lancashire and Fred Hams of Challock, Kent. About 1970, Fred Hams obtained some eggs from Herr Bishof in Germany, and a cockerel from these eggs was crossed with two of Rex Woods' Spanish (± × Minorca) hens he obtained from Fred Tyldesley. Herr Bishof had inherited his stock from his father-in-law whose family had been one of a group of German Spanish fanciers who had been in contact with Bristol fanciers back in the 1880s. Fred Hams obtained more eggs in 1973 and 1974 to build up the quality and numbers of his strain. By 1978 eggs were being sent the other way as the German fanciers had an inbreeding problem. During the 1980s all three of these breeders gave up Spanish as they reduced the size of their collections, but another generation of fanciers, including the author, took over. They remain a minority taste, with no more than ten serious breeder/exhibitors at any one time, but this has included a handful of specialists who have kept them for twenty years or more.

Black Spanish, large cockerel, circa 1923. Note extreme length of white face.

Spanish in the USA

According to John H. Robinson (*Popular Breeds of Domestic Poultry*, 1924), White-Faced Spanish were first seen in America in 1825 by 'John Oldbird'. Hardy, smallish faced, Spanish hens were kept in large numbers during the 1860s in the eastern states as far west as Ohio. There were still a few commercial flocks of Spanish in America as late as 1895, but most had been replaced by Leghorns.

Spanish gradually became a 'high class exhibition breed', 'high class' because only expert specialists could stage winners with extra large, blemish-free, faces. The largest classes were seen at the Boston Show, with most of the exhibitors coming from Massachusetts, New York State and south-eastern parts of Canada. The secretary of the American Spanish Club in the years before the First World War was M.W. Lindsey of Northville, NY.

In October 1912, four Spanish were sent to R.A. Rowan of Los Angeles, California by British fancier Joseph Smith. Their journey was broken at New York, where they won the Spanish classes at a major show. Spanish

Black Spanish, head of large female. Photo: John Tarren

declined in popularity during the 1920s and 1930s in America and Canada as they did in Europe, with interest only growing again in recent times. A new 'Blue Andalusian and White-Faced Spanish Breeders Club' was formed at the American National Show, Ohio, 1998. At the same event, twenty-nine large and eighteen bantam (including one Blue) Spanish were entered along with sixty-two large and fifteen bantam Andalusians. It is not known when the original club closed down.

Spanish in Australia

No details were available at the time of writing about the first importations of Spanish to Australasia. There were many ships going down there from the UK during the nineteenth century, and Spanish fowls were probably on quite a few of them. They declined in popularity there, as elsewhere, during the 1890s. George Woodward contributed an article on the fancy in Australia in the 1911 *FWYB* in which he said that a new Spanish Club had been formed, with most of its eighty-plus members in the state of Victoria. An entry of fifty–sixty birds was being expected at the (then) forthcoming

Black Spanish, large male. Photo: John Tarren

show (1912). This club does not seem to have lasted very long however, and the 1911 article hints at the cause of its fragility. Apparently, the club held a lot of social events, such as concerts, dances and card tournaments. Bear in mind that back in 1911 a lot of people kept a few chickens anyway, and some of the club members may have chosen Spanish just for the socials. Despite these setbacks, there are still some Spanish in Australia in 2005.

Other Colours of Spanish

Like the well-known saying attributed to Henry Ford, Spanish have generally been available in 'any colour, so long as it's black'. A few White-Faced White-Spanish have appeared occasionally, including some that were entered in the 1852 Birmingham Show. It is not uncommon to have Black Spanish with a few white feathers that might be bred together or crossed with White Minorcas to revive this ancient colour variety. They have never appealed to many because the white face does not look right with white plumage. However, as they were known to exist over 150 years ago they are of historic interest. Blue Spanish, most of which were probably really Lavenders have also been recorded at times over the last century. Blues, Whites and, of course, Blacks were all illustrated by Jean Bungartz in 1891 (Leipzig?). The author tried to make a strain of large Blues/Lavenders from a cross with large 'Blue' Minorcas about 1990. A couple of good specimens were produced, but most were either too light in colour or covered in small black specks. As space was limited, they were passed on to two other fanciers, and both of these gave up after a few more years. Barreds were mentioned by Wandelt and Wolters (1996). As the white face has always been the main focus of interest on exhibition Spanish, plumage colours remained a secondary feature.

Spanish Bantam

These have been made several times, only to die out after a few years. It is to be hoped that the current stock lasts longer. W.F. Entwisle's book *Bantams* (1894) mentioned the first-known attempt by James C. Lyell. He started to make them about 1880, but progress was slow because he only had space for a few birds. By 1894 his birds were as small as Black Rosecombs, but the hens had upright combs and both sexes were red between eye and comb.

Mr Thompson was the next Spanish Bantam breeder we know of today. He won first prize in the AOV Bantam Cock class at the 1901 London Dairy (twelve entries in class) and Crystal Palace (eleven), but had to make do with third at the 'International Show' at Alexandra Palace (fifteen), an exhibition/concert hall in London. He was also mentioned in Pringle Proud's book *Bantams as a Hobby* (first edition, 1901) as winning 1st and 2nd in AOV Bantam at

Blue Spanish, large male. Photo: John Tarren. Bred by the author, using Blue Minorcas, this was the only good male in this short-lived experiment.

Blue Spanish, large female. Photo: John Tarren. This was the most evenly coloured specimen ever bred in this strain, most had black specks all over the plumage. Perhaps another fancier might have better luck in the future.

Cliff Lowe, better known in the 1990–2005 period for his Brahma Bantams, also had a go at making Spanish Bantams in the 1950s and 1960s, but they seem not to have interested anyone else. The strain currently seen in the UK, Germany and elsewhere in Europe have all come from Fred Hams in Kent, who, as mentioned above, was also a key breeder in saving large Spanish from extinction. He started this project in 1978 or 1979 by artificially inseminating sperm from large Spanish males into three different bantam hens: a Black Dutch, a Black Rosecomb with a partly white face and a tiny Minorca Bantam hen he obtained from the Netherlands. He tried other breeds over the next few years, probably including Black OEG, and certainly including an accidental mating with a Black German Langshan Bantam. Artificial insemination was only needed in the first year, but some of the later stages of the process involved intensive selection. In 1983 only thirty birds were

Kendal when Mr Proud was judging. Kendal, in the Lake District area of north-west England, was Mr Thompson's local show.

There then follows a twenty year gap before the next definite mention of anyone having Spanish Bantams, and that is just an advert in the 1921 FWYB put in by Mr J.M. Cooper of 5 Gordon Road, Herne Bay, Kent. He also bred large Spanish, and it is interesting to note he lived at the opposite corner of England from Mr Thompson.

From then until the 1960s, Spanish Bantams were only spoken of in the past tense, with most commentators wondering why they had not been more popular as the few that these three gentlemen exhibited were much admired. For example, C.A. House (*Bantams and How to Keep Them*, circa 1930) wrote: 'A few years ago several very nice Spanish Bantams made their appearance in the show pen, birds that were indeed miniature Spanish, possessing beautiful long, smooth, open, white faces, grand shape, and beautiful colour and feather.'

Black Spanish, bantam male. Photo: John Tarren

*Black Spanish, bantam female.
Photo: John Tarren*

retained from the 180 bred. Fred first exhibited his new creations in 1984, but a few years later he had passed them all on for others to continue. At this stage, they were small and had completely white faces, but not much length of face; their facial skin was still too hard and lobe-like, not loose and floppy as it should be.

In the twenty years since then, the author and other breeders in the UK and elsewhere in Europe have been gradually refining them. Increasing the size of the face, eradicating the harder tissue that gives the impression of a distinct lobe as opposed to the smooth face that is required, and reducing their overall body size that is helped by selection for tighter plumage.

Spanish Bantams were accepted to the American Standard of Perfection in 1960 as a result of a strain being made by Joe L. Templeton of Derby, Kansas. He described the process of making them in the 1963 *A.B.A. Yearbook*, that was repeated in Fred Jeffrey's books (*Bantam Chickens* in 1974, *Bantam Breeding and Genetics* in 1977). Mr Templeton started in 1952 with a large Spanish × Black Rosecomb Bantam mating, with Black Leghorn Bantams being added later. In 1959 another approach was tried, large Spanish × Black OEG, these being some hens that the breeder was glad to get rid of as they had a serious fault for OEG, white in face. It is not known (by the author) if the Spanish bantams shown at Ohio in 1998 were directly from Mr Templeton's strain.

Description

In British Poultry Standards the Scale of Points for Spanish gives (out of 100) thirty-five points for the face and fifteen points for comb and wattles. Thus half, 50/100, of the official judging points are for the head. All of the facial skin should be pure white without red patches or brownish scabs or other blemishes. They are most likely to be red between the eye and the comb or between the wattles. Reddening over a wider area is caused by prolonged exposure to strong sunlight, wind or rain. Spots and blisters can be caused by dirt, scratches, fighting or hen pecking. Some males have had spectacularly long, white faces in the past, and large Spanish have revived enough in recent years for birds equal to the best of the 1900–25 era to be a possibility once again in the not too distant future. The skin should hang loosely down with no ridges to suggest a separate lobe structure. Although a smooth face is desirable, some folding is inevitable on older birds, and most Spanish enthusiasts would rather see a large folded face than a smaller smooth face. Spanish pullets seldom have large faces, but by the time they are three years old some can be really impressive. Aim for both sexes to 'have the white below the red', that is, the face to be longer than the wattles. Their single combs should be upright on males, flopping over on females. The combs of winning males should not be too large, as a lower comb makes the white face look longer; an optical illusion, but it works. Specialists might consider double mating for comb formation – larger combed cocks will give floppier combed hens. Eyes should be as dark as possible, but not if this brings dark facial skin pigmentation, as sometimes happens.

General body size and shape is typically Mediterranean, perhaps longer in neck and leg than related breeds. Tails are quite large and fanned. Body plumage is 'short and close'. A slight change of perspective is needed when selecting Spanish Bantams as long-backed, long-necked and long-legged specimens are simply too big to be acceptable bantams. Very few are as small as we would like, the ideal being about the same as Andalusian Bantams. Shanks and feet should be pale slate blue. Large Spanish sometimes have nearly white shanks, which is a fault, and Spanish Bantams usually have nearly black shanks and feet, which is also wrong. Large Spanish males should weigh 3.2kg (7lb), females 2.7kg (6lb). Only a few do, so do not panic. Current British Standards specify 1,075g (38oz) males and 910g (32oz) females for the bantams, which is probably what most of them are. A reduction to 900g and 800g should be aimed for, preferably even lower.

Special Management Requirements

Spanish are demanding birds if they are to be kept in show winning condition. They are best kept by (near) specialists, that is, those with not too many other breeds. Victorian fanciers would spend hours gently massaging the faces of their show birds to obtain maximum size and smooth out any folds. No one is likely to have the time or inclination to do this today. They will still pluck out some of the tiny feathers that grow out of the white skin, wash and baby powder faces before showing.

In order to maintain a prize-winning flock of this rare breed over an extended period, a considerable number should be kept. Spanish males, especially large fowl, are susceptible to extreme cold, facial skin blemishes and having their faces pecked by their hens – for all these reasons, as many males should be kept as possible. Hen pecking can be minimized by having at least two males for each pen of hens, and alternating them each week. After a week in a pen on his own, the 'rested' males will not allow hens to peck him. After a while the males seem to get lethargic. Time to swap them again. Therefore a number of single bird pens are needed.

Outside runs should be protected from the wind, rain and extreme sunlight. Remember that the two traditional strongholds were London and Bristol in the milder south, and most of them lived in the city centres, suburbs or sheltered nearby villages.

Spanish Bantam chicks are usually easy to rear, but the large ones are slow to develop, slow to feather, and the black feathering is prone to 'fretting', especially in the tail. As said above, not many are up to the ideal standard weights. For all these reasons, they should be fed on the best available chick crumbs, possibly even turkey starter crumbs, for much longer than the eight weeks that is adequate for quicker developing breeds.

STARUMSE RONDKAMM

Virtually no information has been found about this rare Dutch breed that has a large, almost circular (that is, with no rear spike or leader) rose comb. This feature suggests they are related to Assendelftse Hoenders and Derbyshire Redcaps.

THÜLRINGER BARTHUHN/ THÜRINGIAN BEARDED ('Thüringer Pausbäckchen' until 1907)

The old name, *Pausbäckchen* roughly translates as a German equivalent to the English affectionate terms 'chubby cheeks' or 'mutton chops'. They were standardized on 8 March 1907 under their present name, largely as a result of two breeders from the town of Ruhla.

They are much more ancient than this however, the first recorded mention being by J.M. Bechstein in 1793. It is thought that they may be related to Nederlandse Uilebaarden (Dutch Owlbeards). Although there are more colour varieties now, for most of their history there were only two, Gold and Silver Spangled. Generally, they are typical, small, active cottagers' fowls. The beards and spangled plumage pattern is almost certainly the result of hundreds of farmers' wives deciding to keep the pretty cockerels for breeding, with their duller brothers being eaten. 'Pausbäckchens' were, like hobbyists' birds today, pets as well as egg and meat producers. When poultry showing started in Germany in the late nineteenth century, they soon became a great favourite, although some of the newer colour varieties have remained rare. A few have been seen at British shows since 2000.

Thüringer Zwerg-Barthuhn/ Thüringian Bearded Bantam

The miniature version of this breed was first developed in the 1890s, but current strains were probably from post-1945 remakes. Ernst Florschütz is mentioned as one of the pioneers in the 1890–1914 period. Many breeds have been used to make them, as appropriate for each colour variety. Large Thüringians have been crossed with Barbu d'Anvers, Chabos, Deutsche Zwerghuhn (German Bantams), Spangled Hamburgh Bantams and Welsummer Bantams.

Some Thüringian Bearded Bantams have also been seen at British shows since 2000, and out of all the colours, in both sizes, the Chamois, or White Spangled Buff Thüringian Bantams seem most likely to attract sustained interest.

Description

The general impression is of a compact light breed with a broad, rounded breast. Most of the parts of the body: back, legs, neck and so on are described as 'medium length'. Large fowl weights range from 2kg (4½lb) pullets to 3kg (6½lb) adult cocks. The miniatures are a small 600–700g (21–24oz).

Head points are very important, the two main features are fairly small, straight single combs and thick, bushy beards. Although thick, their beards are not particularly long, and are not in the tri-lobed formation required on Faverolles and some other breeds.

Some of the colour varieties do not require further explanation: Barred, Black, Blue, Buff, White and Black-Red/Partridge. The three spangled varieties: Chamois, Gold and Silver are more popular and have quite large, but not overlapping, spangles. The main tail feathers, beards and lower parts, belly and over the hock joints are more or less solid black (Golds and Silvers) or white (Chamois). Some, especially Silver females, have spangled tail feathers (as Hamburghs) that no one seems to regard as a fault.

UKRAINSKAJA USCHANKA/ UKRAINIAN USCHANKA

In general appearance these look like a smaller version of an Orloff. They were recorded back in the seventeenth century, and were made known to Western poultry people by the much travelled Edward Brown in the 1890s. He said they had single combs, but Wandelt and Wolters (1996) describe and show a photo of a rosecombed breed. Uschankas are bearded. Known colour varieties include: Barred/Cuckoo, Black, Red, Spangled, White and dark Black-Reds (crow-winged males and red-hackled black females).

European Light Breeds

Gold Spangled Thüringer Barthuhn, large male. Artist: August Rixen

Gold Spangled Thüringer Barthuhn, large female. Artist: August Rixen

As far as is known none have ever been brought over to the UK although a trio might have been exhibited at the 1930 World Poultry Congress at Crystal Palace, London. Since the fall of Communism, with the resulting flow of people from former Soviet-bloc countries now living in Britain and the rest of the EU, it is only a matter of time before some hatching eggs arrive.

VOGTLÄNDER

German breeder Martin Neudeck (1889–1980) started to make his breed in the 1920s, but for reasons unknown to the author they were not recognized until 1973. Breeds mentioned by Wandelt and Wolters (1996) in their make-up included Andalusians, Dominiques, Rheinlanders and Welsummers. A '1973 breed' would not have been included in *Rare Poultry Breeds*, but as Mr Neudeck started in 1920, they are. The author does not know how many still exist.

The result of his efforts is a tight-feathered, athletic-looking breed with a compact, smooth-topped rose comb (as some German type Wyandottes), white ear lobes and long wattles. The birds illustrated in Wandelt and Wolters' book are red-hackled blues.

VORWERKHUHN/VORWERK

This is another breed that was very much one man's idea – Oskar Vorwerk from Hamburg-Othmarschen. He started in 1902 to make his concept a reality, this being a more useful version of the Lakenfelder with the white (that shows the dirt) being replaced with buff. He first exhibited them in the new varieties section of the 1912 show and they were fully recognized in 1913 or 1919 (accounts vary). Some were brought over to England for a major 'International Show' held in Essex in 1935, where they were called 'Buff Lakenfelders'. No records have been found of any others in the UK until recent times (*see below*).

Breeds used to make them included: Lakenfelders, Andalusians, utility-type Buff Orpingtons and Buff Ramelslohers (then called 'Hittfeldern', for reasons unknown). Poultrykeeping in Germany was not very developed in the 1902–14 period, or for some time thereafter. Oskar Vorwerk's idea of a dual-purpose smallholders' breed suited their needs, unlike in America where the beginnings of today's separate egg and poultry meat sectors, both run on industrial lines, had already started to appear.

As with all German breeds, fanciers had to rebuild strains from the survivors and by crossing with other suitable breeds. In the case of the Vorwerk, this happened separately in both East and West Germany. By 1960 large numbers could be seen at shows in both the DDR and FDR.

They next appeared in Britain in the late 1970s, imported by Jane Wallis, of Arundel, Sussex who started 'The Vorwerk Club of Great Britain' in 1980 instead of joining in with the Rare Poultry Society. The breed did not attract enough interest for a separate club to be viable, so Vorwerks came into the RPS in 1984.

Zwerg-Vorwerkhuhn/ Vorwerk Bantam

Breeders started to make a miniature version in the 1920s, not long after the original large fowl appeared. They did not really become established until the mid-1950s because of the war. Then they were mainly an East German production. Breeds used were undersized large Vorwerks, Buff Orpington Bantams, Gold Duckwing German Bantams and Barnevelder Bantams. This mixture, after several decades of selective breeding, is smaller than one would expect from the breeds used (only German bantams are small). Those that combine small size with correct markings are very pretty and are bound to appeal to many.

Description

They are a compact utility breed. Surplus cockerels make good quality, traditional-type table birds, fine-boned and full of flavour. Not very large, but they can be thought of as the German equivalent of the famous French La Bresse.

When standing normally, the back should be slightly sloping and straight, with no

Vorwek, large male. Photo: John Tarren

Vorwek, large female. Photo: John Tarren

cushion effect on the females. Their tails are a little shorter, lower and more folded (whipped) than those of Lakenfelders. They are fuller in the body than Lakenfelders as well. German Standards require 3–3.5kg (6½–7½lb) large males, 2–2.5kg (4½–5½lb) large females, 900g (32oz) bantam males and 800g (28oz) bantam females.

Combs are single, medium or smallish in size and straight. They are allowed to fall over slightly at the rear on hens. Wattles are average and the white lobes are quite small. The beak should be horn-coloured or bluish, to harmonize with the slate shanks and feet.

The plumage pattern is easy to describe, but very difficult to breed. Head, neck and tail should be as near solid black as possible, plus some black markings on wing feathers. Body plumage should be a deep buff with grey under-colour. Males have some black striping in saddle hackles. Most birds with solid black necks will have some black speckles on the parts that should be buff. Most birds with clear buff bodies will have some buff in neck feathering. Winning show birds have usually been improved by careful removal of quite a lot of wrong coloured feathers. Unfortunately, a lot have so many wrong-coloured feathers that it would be cruel to do it, and would be too obvious. Since 2000 a very pretty blue and buff variety has appeared in bantam form. The only other genetically possible colour variation of this pattern would be lavender and cream.

WESTFÄLLISCHE TOTLEGER

The second part of this breed's name roughly translates as 'ultimate layer'. In truth they are fairly good layers, but not that good. They look like a cross between Ostfriesischen Möwen and Brakels, and may have been exactly that, from crosses dating from about 1850. Or they may have simply been a tidied-up version of indigenous chickens of the Nordrhein-Westfalen region. The breed was standardized in 1904, and the specialist breed club was formed in 1920. It became

Vorwek, bantam female. Photo: John Tarren

quite popular in Germany, but virtually unknown elsewhere. Entries at major German shows since 1980, between twenty and sixty birds, suggest moderate support.

Their general appearance would make some poultry fanciers think of a rosecomb version of Brakels. Westfällische Totlegers have low, smooth-topped combs with leaders that follow the line of the skull. Ear lobes are small, round and bluish-white. Eyes are dark brown and shanks are slate-blue. Standard weights are quite low, ranging from 1.5kg (3¼lb) pullets up to 2.5kg (5½lb) adult cocks. This probably explains why no bantam version has been produced.

Two colour varieties are recognized, Gold and Silver, the latter being most popular. The pattern is virtually the same as that of Brakels, autosomal barring all over except for clear silver or gold hackles, just the neck of females, neck and saddle of males.

ZOTTEGEMS HOEN

These are essentially a colour variety of the Brakel in all but name, and were listed as such by Edward Brown in 1905 (*Races of Domestic Poultry*) and 1929 (*Poultry Breeding and Production*). It is not clear when they were defined as a separate breed locally, possibly around 1912. Brakel is a village, and Zottegem a small town, both being south of Gent (Ghent) in Belgium. Traditionally, poultry and eggs in this area supplied Gent and Brussels, many via the market town of Aalst.

Edward Brown described them as 'Black-Headed Brakels', which sums up the two colour varieties, as normal Gold and Silver Brakels except that the clear gold or silver neck and saddle hackles are replaced with black feathering. The head feathering is solid black, but down the neck and on the males' saddles the feathers have white underfluff, giving a washed-out black effect.

It is thought they were originally produced from crosses between normal Brakels and Black-Headed Dutch Owlbeards. It is not certain if Zottegems Hoenders have existed continually, or if old strains died out and were remade in recent decades. The only record the author had at the time of writing about them was the 1983 Aalst Show catalogue, where Robert Van Den Berghe of Zwalm entered six Silver and three Gold 'Zottegemse Zwart Kap'. If a new strain ever needs to be made, crosses between normal Brakels and Lakenvelders or Vorwerks might be useful as Black-Headed Owlbeards (Moorkop Nederlandse Uilebaarden) are rare.

The remaining plumage is barred, as on Brakels. All other details: body shape, eye, lobe, beak, shank and feet colour are also as for Brakels.

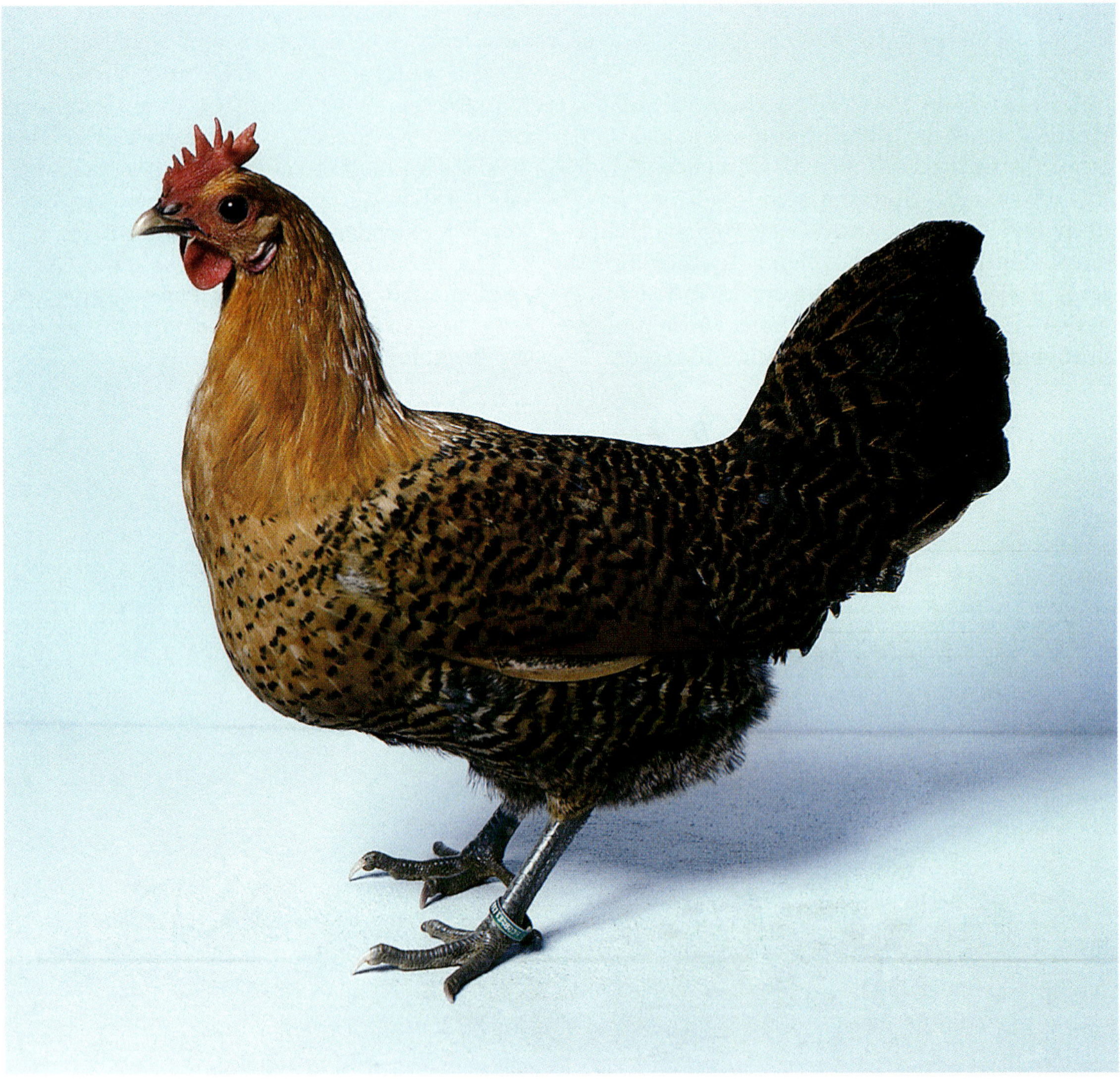

CHAPTER 2

Egyptian Breeds

An Egyptian breed, the Fayoumi, was imported by the Domestic Fowl Trust to England in 1984, and since then have been exhibited here. Although Fayoumi Fowl are very similar in general appearance to many in the European Light Breeds chapter, Egypt is in Africa, not Europe, so they are (almost) in a chapter of their own. This book is intended for hobbyist poultry keepers, mostly in those countries where poultry shows are regularly held. Therefore the emphasis has been on those breeds that might be of interest and available to these people. As evident from the brevity of this chapter, and the absence of a broader 'African Breeds' chapter, neither the birds, nor information about them, are readily available at present.

Although not very evident in recent times, large-scale artificial incubation and egg production began in Egypt some 4,000 years ago. The operators lived in the incubators, that held over 10,000 eggs. They had to, as there were no thermometers then, and only by living in the incubater that was also the operator's house, did they remain sensitive to slight changes in temperature. The author does not know how long they existed.

BALADY

This is the name given to the small, 'jungle fowl size' chickens to be seen, in previous decades at least, running around many villages in Egypt. As far as is known, and this

Silver Pencilled Fayoumi, large male. Photo: John Tarren

is a bit limited, most are or were small single-combed chickens of various colours, most probably variations of Black-Red plumage pattern. They may be much rarer now, as most commercial producers will probably be using hybrid broilers, many of which are the Naked-Neck type so popular now.

Egyptian Breeds

Silver Pencilled Fayoumi, large female. Photo: John Tarren

BIGAWI

Larger than Balady, and as Fayoumi, they are also from the area around the city of El Faiyum, south-west of El Qahira (Cairo). Bigawi may possibly be a different name for Fayoumi.

FAYOUMI

They are a light and very active breed, no doubt the result of the forces of 'survival of the fittest' over some 4,000 years in Egypt. Fayoumi are quick to feather and mature, good fliers and 'screamers' when handled. Obviously they are adapted to living in the hot, dry environment of Egypt. None of these qualities are likely to be tested, nor are they particularly desirable in English fanciers' gardens; and their wildness will not be appreciated by many show judges.

Two colour plumage colour varieties have been standardized in the UK, Gold and Silver Pencilled. Apparently these represent the middle of a wider range of markings from something similar to the Columbian pattern at the light end, to others with much heavier black markings.

It is to be hoped that Fayoumi continue to thrive in Egypt, and that a few flocks remain in the UK and elswhere around the world as 'insurance', in case of disaster, however this breed is unlikely to become a common sight at the shows.

Description

Fayoumi are alert and very active, standing well up and with high tail carriage. They are a small breed, standard weights ranging from 1kg (2¼lb) pullets to 1.8kg (4lb) adult cocks. This does not leave much scope for a bantam version. They have small single combs, red lobes, dark brown eyes and slate-blue shanks/feet.

Silver males have white head, neck, breast, shoulder and saddle feathering. Their tail

Gold Pencilled Fayoumi, large female. Photo: John Tarren

feathers are black, with some white edging. They have black barring on the thighs, belly and some wing feathers. Females have the same barring over most of their body apart from a clear white head and neck with the barring on the breast being very sparse and irregular. Gold Fayoumis have the same black markings on a gold ground colour.

Special Management Requirements
Housing should allow for their active temperament and tendency to fly. If it is intended to show them, growing birds should be regularly handled and petted to tame them.

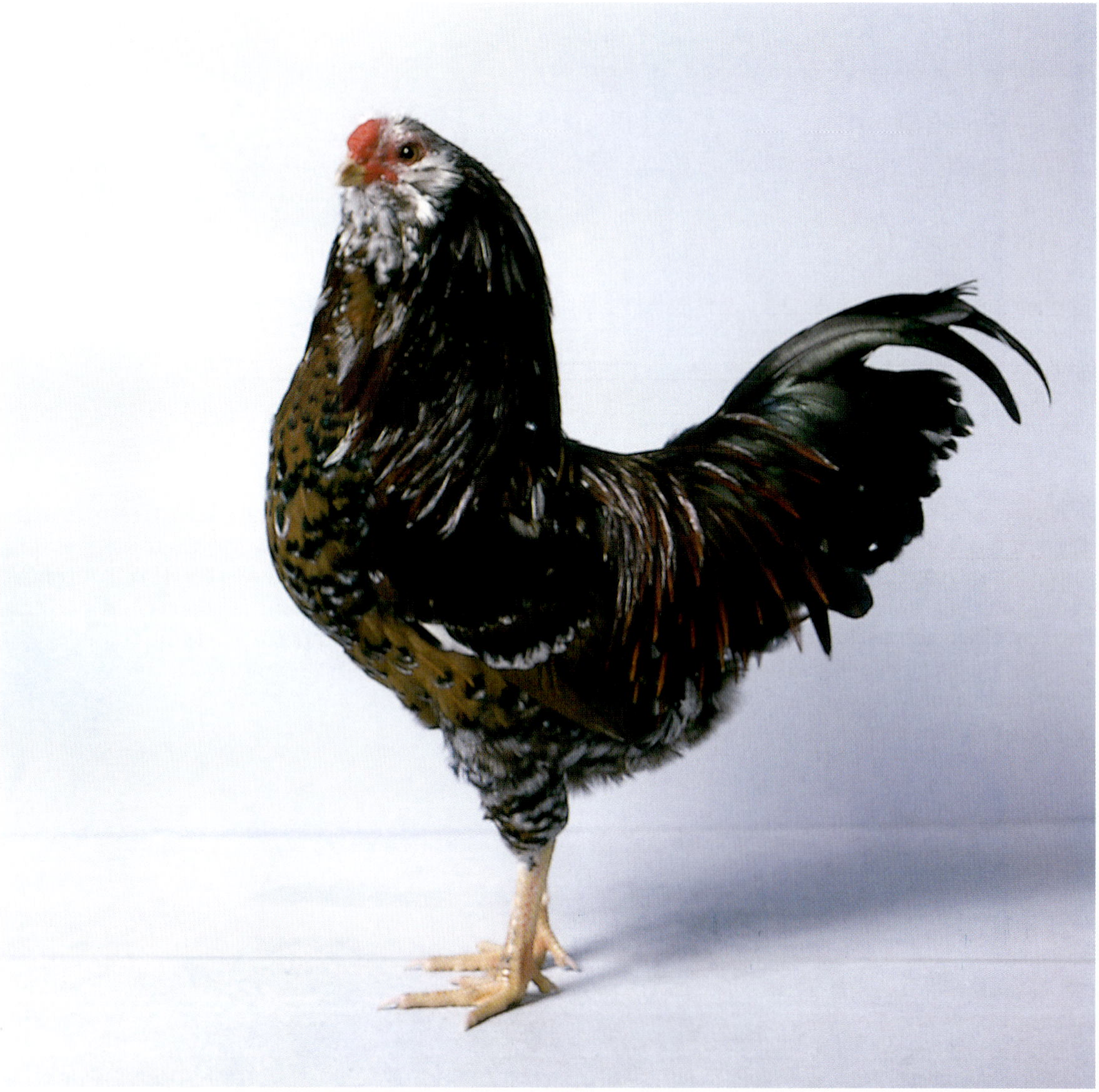

CHAPTER 3

Long-Tailed and Game-Related Breeds

This book does not generally cover the Game Fowl breeds, partly because many books have been published on these breeds over the last century already; partly because the history of their development could not be given without a lot of potentially controversial detail about cockfighting.

Similar breeds in other chapters of this book:

Rumpless Game ('Game-Related') in Rumpless Breeds chapter
Koeyoshi ('Game-Related'), Totenko ('Long-Tailed') in Long-Crowing Breeds, Chapter 5

Breeds in this chapter, and their countries of origin:

Cubalaya – Cuba
Minohiki – Japan
O-Naga-Dori – Japan ('proper' Japanese long-tails)
Orioff – Iran, then popularized in Russia
Phoenix – European modified version of O-Naga-Dori
Satsumadori – Japan
Shokoku – Japan
Sumatra – stock imported from Sumatra, refined/standardized in UK, USA and Germany,
Twentse Hoen/Kraienköppe – Netherlands and Germany
Yokohama – Germany (from imported Minohiki and O-Naga-Dori?)

CUBALAYA

This breed first came to wider notice at the World Poultry Congress, Cleveland, Ohio, USA, 28 July–7 August, 1939. The author has not found many details of their origin, although Loyl Stromberg (*Poultry of the World*, 1996) said the main breeder was Sr Domingo Montane, circa 1880–1920. The breeds from which he (and/or others?) probably used to make Cubalayas included: Spanish Game, American Pit Game, large Orientals (native Malays? Shamo?) and long-tailed Orientals (Philippine variations of Sumatras?).

To give a little historical perspective, it must be remembered that by the late nineteenth century virtually all that was left of the once (before about 1810) vast Spanish empire was Cuba and Puerto Rico in the Americas and The Philippines in Asia. Even these colonies were lost in 1898, after the Spanish-American War, when they effectively came under American control.

Cockfighting was a very popular activity in large parts of the world. In the Americas, each community (most of which were originally immigrants from elsewhere) had brought their own breeds, Game Fowl from Belgium, England, France, Ireland and Spain in the European tradition and a similar range from Asia, ranging from heavyweights like Shamo, smaller 'naked heelers' like some strains of Asil and high-flying, long-tailed Game, like the birds that were later modified into the exhibition Sumatra by British and American show fanciers. Some breeders then based in the Americas (from the USA down to Argentina) kept to whichever breed they had been used to, but others tried crossing with whichever (to them) exotic new breeds came their way.

The obvious book to consult for a detailed account of the origins of any Game Fowl breed in the Americas is C.A. Finsterbusch's *Cockfighting All Over the World* (1929, published USA, although the author was Chilean). However, there are virtually no details given, beyond saying that Cubans were more interested in 'naked heel' cockfights than those with steel spurs. This suggests that they would have kept large Orientals (like Shamo) rather than smaller breeds like Spanish Game. He did not mention Cubalayas, or any similar long-tailed type in Cuba; and Mr Finsterbusch was obsessively interested in Game breeds. This suggests that Cubalayas were either very new, very rare, or Sr Montane was intending to create an ornamental breed. If they had been used in Cuban cockpits, Finsterbusch would have written about them.

The first recorded American importer was Horst W. Schmudde, a prominent exhibitor of several Oriental breeds. They were first exhibited at the live poultry show that was held in conjunction with the Seventh World Poultry Congress at Cleveland, Ohio, 28 July–7 August, 1939. Unfortunately, no one contributed to the published proceedings about them. There is a Cubalaya Club for them in the USA. Some were sent to a German fancier, Franz Swist of Bremen in 1978. They were accepted to German Standards in 1983. None had been seen at time of writing at any UK shows. Cubalayas were still being bred in Cuba at least as recently as 1980.

There are three standardized colour varieties of Cubalayas, Black-Breasted Red, White and Black. Other colours have been known. The author (2005) is not aware of any in Europe.

Cubalaya Bantam

These were added to American Standards in 1960 in the same three colours as the large fowl version. One American breeder, Merle L. Keaffaber, made a strain, not the first, in the mid-sixties. Before using the obvious undersized large Cubalaya pullets, he had made a roughly 'lookalike' bantam using Black-Breasted Red Malay, Modern and Rosecomb Bantams.

Description

In general appearance, Cubalayas are broadly similar to Sumatras, with a tail that is long, low and fanned. In the latter respect, they have similarities to Satsumadori. There is more obvious difference between Cubalayas and Sumatras or Satsumadori in the female's tail carriage. Cubalaya females' tails sweep straight on from the back down to the ground. Large fowl weigh from 1.5kg (3¼lb) pullets up to 2.5kg (5½lb) adult cocks, with the equivalent bantam weights of 570g (20oz) and 740g (26oz).

They have typical Oriental breed heads with a short thick beak and compact pea comb. Another feature that helps to distinguish them from Sumatras, is that all colour varieties of Cubalayas, especially including Blacks, have red comb, face, wattles and lobes. Eye colour is described as 'reddish bay', contrasting with the very dark eyes of Sumatras and rather lighter eye colour of Satsumadori. Shanks and feet are white on White and Black-Breasted Red Cubalayas and slate on Blacks.

Black and White Cubalayas do not need further plumage colour description. Black-Breasted Red males are a bright shade of golden-red in top colour. The females are Cinnamon, like many Malay hens.

Special Management Requirements

Cubalayas have tails that are long enough to touch the ground, but not so long that they require a lot of very specialized management. Houses should be large and the high perches be placed well out from the walls.

MINOHIKI

They are believed to have been developed in Japan from birds originally imported from somewhere in China. Once in Japan, they were mainly bred in the Prefectures Aichi and Shizuoka on the main Japanese island of Honshu, between Kyoto and Tokyo. 'Mino' can be translated as saddle feather and 'Hiki' as dragging. The tails on males also drag along the ground, but no more than those of other long-tailed breeds normally seen at European or American shows. Minohiki are probably the nearest to the original type of all the long-tailed breeds. At the time of writing (2005) the author was not aware of any Minohiki in Europe.

Ginger Minohiki, male. The property of Mr Yagi, Japan. Photo: Julia Keeling

Description

Most Minohikis have walnut combs, although pea combs are also permitted. They have long neck, saddle and tail feathers, but their body feathering is quite tight. Tail feathers on adult males are typically about 90cm (35in) long. As might be expected with breeds that are probably related to Shamo-like breeds, they have light-coloured eyes. Weights range from 1.35kg (3lb) pullets up to 2.5kg (5½lb) adult cocks. The main colour varieties are Black-Red/Partridge, Silver Duckwing and Black-Tailed Orange-red.

Ginger Minohiki at Shizuoka Show, Japan. Photo: Julia Keeling

O-NAGA-DORI

Many of the standard poultry books include descriptions of O-Naga-Dori, often with photos of fantastic specimens with spectacular tails. However, the best account of their history and the special techniques necessary to keep them in good order remains an excellent article by Frank X. Ogasawara Ph.D in the December 1970 issue of *National Geographic Magazine*.

As might be guessed, the very long-tailed O-Naga-Dori started from now unrecorded matings of the more ancient, moderately long-tailed, breeds: Minohiki, Shôkoku and Totenko. Formal ceremonies played an important part in seventeenth-century Japanese society. Some events involved processions in which many banners, decorated spears and other ornamental weapons were carried. Tail feathers from the above breeds, up to a metre long, were among the decorative items used. Lord Yamanouchi, second Tozama Daimyo of the Tosa region around the city of Kochi on the Japanese island of Shikoku wished to have the best possible processions, so put an expert poultry breeder in his domain, Riemon Takechi, to work on increasing tail length. Yamanouchi was an 'outside' (Tozama) 'lord' (Daimyo), which means he was one of quite a few lords who held their lands more or less independently of the Emperor, and was not entirely trusted by the Tokugawa (central) administration. The Tozama Daimyo were kept loyal (?) by the Emperor by several means, including periods when the families of the lords were required to stay at the then capital, Kyoto. They may have been treated as honoured members of the court, but were really hostages. With this atmosphere of mutual suspicion, it is quite understandable why Lord Yamanouchi would have been anxious to put on a good show on all occasions involving the Emperor or his representatives.

Dr Ogasawara mentioned that there was (by 1970) a statue near the railway line between Kochi and Nankoku dedicated to Riemon Takechi. It is pleasing to learn that in these more egalitarian days, it was Riemon Takechi, the man who actually did the selective breeding, who is celebrated in stone, not the lord who ordered him to do this specialized work.

Riemon Takechi was mentioned in the article as being active in 'about 1655', but no further dates are given. It is possible that after his death (date unknown), his work may have been continued by his son or some other relation. The great difference between O-Naga-Dori and the other long-tailed breeds is that O-Naga-Dori do not drop their tail feathers in an annual moult; instead the feathers just keep growing. It is not known when this mutation occurred.

Japan's closed feudal system continued until the mid-nineteenth century; when pressures from both internal dissidents and outside (the European powers and the USA) caused Japan to start modernizing and open up to the rest of the world. Although ancient traditions continued to be cherished by many Japanese, the number of ceremonial processions, and therefore the demand for long feathers, declined. However, by this time the breed had attracted a wider interest in Japanese society. The keeping of O-Naga-Dori became an absorbing hobby for anyone with the time and space for it, alongside Bonsai trees and Koi carp. An association of O-Naga-Dori breeders was formed in 1908 that was boosted by the Japanese Government recognizing the breed's historical and cultural significance in 1923. Enough birds survived the Second World War for the association to be reactivated, and the official support reaffirmed, in 1952.

The breeders association no doubt held meetings and competitions, but no details were available to the author at the time of writing. Some breeders earned a living from their birds by opening to the public as a tourist attraction. Picture postcards, mostly of white cocks, were photographed (in monochrome), some of which advertise: 'When in Miyanoshita, be sure to visit Yamatoya's Art Gallery, and see our famous Long-Tailed Fowls.' Another breeder, also in Miyanoshita, was S.M. Shiba. His collection ('near Fujiya Hotel') included a cock that had a 20ft (6m) long tail when it was five years old. Mr Kubota, one of the breeders featured in the *National Geographic* article, carried out some public speaking engagements, but Dr Ogasawara was obviously far too polite to ask anything about money earned from them.

Top quality O-Naga-Dori are rare all over the world – there are not even that many in Japan. Anyone who has ever seriously thought about keeping them has studied the techniques necessary to maintain really long tails. After doing so, most have given up the idea, perhaps deciding to keep one of the other, less extreme, breeds in this chapter instead. The limited number of existing enthusiasts will probably have some stock available to sell to those who would like to try them, and have a track record of success with other breeds to prove their competence. Valuable stock should not be wasted on passing whims.

There are no O-Naga-Dori Bantams, Phoenix Bantams being the nearest available.

Description

Long tail and saddle feathers are obviously the main feature. The record holder at one time had a 12m (39ft) long tail. Cocks with tails over 7m (23ft) will probably be quite old, almost geriatric in chicken terms. The feath-

Long-Tailed and Game-Related Breeds

Japanese long-tailed fowls have long been a favourite with artists.

White O-Naga-Dori, male. One of a number of postcards produced by breeders in Japan who had their birds on display as a tourist attraction during the 1920s and 1930s.

ers will be many years old as well, and very few birds or feathers last that long. Most breeders will think they have done well if they have cocks with tails over 3m (10ft) long; very well indeed with a 6m (20ft) plus tail.

They have single combs, white ear lobes and yellow shanks/feet. The three main colour varieties are Black-Red/Partridge, Silver Duckwing and White. There are two additional, rarer colours: Golden Duckwing and Gingers (Black-Tailed Orange-Red).

Special Management Requirements

Those cockerels that are selected to become show birds are kept in very confined, high level cages, with the tail hanging out of the back. They are taken 'walkies' several times a day, on a lead, with their tails held up by their owner on some kind of frame. There is more to it than that, for example, the buildings in which the show cock cages are kept have to be in a constant, warm temperature to encourage maximum feather growth. A special feeding regime is probably also needed, but no details were available. New breeders will need to spend a lot of time with

The O-Naga-Dori Centre, Kochi, Japan. Photo: Julia Keeling

experienced fanciers to learn all the techniques. One less thing to worry about, it is normal practice to trim the tails of breeding cocks to a manageable length.

ORLOFF (ORLOWSKIE)

Although they have long been called 'Russian Orloffs', this breed originally came from the Gilan region of northern Iran, where they were named 'Chilianskaias'. They were 'discovered' (as if the people who had been keeping them for generations did not count) and taken to Russia by Grand Admiral Count Alexis Grigorievich Orlov (or Orloff, as usually named in English). He was born in 1737, had an exciting period in political intrigue and military campaigns until the mid-1770s (consult a book on Russian history for full details) after which he settled down to a quieter life devoted to his interests in natural history and horse breeding. He had created the Orlov Trotter breed of horse before he died in 1808.

No information has been found by the author on the origin of the Orlov breed either in Iran, or later in Russia until it became known first to the German, and then British and other Western poultry experts in the late nineteenth century. A party of German and Austrian poultry experts saw the birds when touring Russia in 1880, and they came to wider notice at the 1899 International Poultry Exhibition at St Petersburg. British poultry expert, Edward Brown (knighted Sir Edward in 1930) went to the St Petersburg Show, and brought a trio of Spangled Orloffs back with him. He assumed that they were the first Orloffs brought to the UK, but an article by Rhys Llewellyn in the 1935 *Feathered World Yearbook* suggests that they were known in Scotland back in the 1870s. Unfortunately, both of the hens in Brown's trio died before he was able to hatch any eggs.

Details of the next importations to Britain have not been found, but it was probably about 1910, and according to *Hick's Encyclopedia of Poultry* they were first exhibited here in 1912. One importation, direct from Russia, was made by a Captain Lawrence. Most of these were of the Mahogany colour variety, and the great contrast between the very glossy males and softer, duller-coloured hens was commented upon. British breeders were informed that the Russian experts concentrated on head character, and virtually ignored plumage colour.

They created a flurry of interest, and so the British Orloff Club was formed in 1915. Mrs Colbeck of Boyle Hall, near Wakefield, Yorkshire was the club's Secretary until 1939, when the club's activities were suspended, never to be revived. Most of the breeders, and therefore most of the Orloff Club Shows were held at shows in Scotland or the north of England rather than the Crystal Palace Show favoured by most other breed clubs. Orloffs had a very short period of popularity in the UK, so full details can be given (all are Orloff Club Annual Shows):

30 December 1921, Sheffield, 4 classes, total of 64 entries
?? November 1923, Galashiels, 4 classes, total of 78 entries
21 November 1925, Galashiels, 4 classes, total of 33 entries

In 1925 the Galashiels Show almost clashed with Crystal Palace, so southern Orloff breeders decided not to travel up to Scotland the day after Crystal Palace finished. The southern breeders were able to discuss Orloffs with overseas visitors to Crystal Palace, from all over the world, the most important of which was Dr. Trubenbach, a renowned German expert. He said that although there were many more Orloff breeders in Germany, the British-bred Orloffs were better than most in Germany. No further details of Orloff Club Show entries were available, but it can safely be assumed that displays of about thirty would have been typical.

In 1926, Prince Orloff, whose family was presumably living in exile after the Russian Revolution, bought some Orloffs in the UK

Long-Tailed and Game-Related Breeds

Spangled Orloffs, pair. Artist: Kurt Zander

(Mrs Colbeck?). One of them won the Prix d'Honneur and Medal for Best Orloff at Lille Show. They were Spangles, and presumably the Prince decided to keep them as a momento of his famous ancestor and the family estates that he lost.

Despite their reputation for withstanding harsh climates, Orloffs declined during the hard economic times before the Second World War. Rhode Island Reds outperformed Orloffs as layers and table birds, even in Scotland and the coldest parts of England. It is not known if any Orloffs survived in Britain through the war. All known Orloffs in the UK since the 1970s were from Germany, The Netherlands or direct from Russia.

A party of British poultry fanciers visited the Hanover Show in 1971, where among many other breeds of which their generation had only ever seen pictures, were fifty-one large Orloffs, and an unrecorded number of Orloff Bantams. There were thirty-three Spangleds, eleven Mahoganies and seven Whites. Major German shows have consistently had comparable entries of Orloffs. At time of writing (2005) it was unlikely that there were fifty-one large Orloffs in the whole of Britain.

Orloff Bantam

These were first made in Germany about 1920, but because of world events did not progress much until after 1945. A few of the original strain had survived, and by crossing them with Rhode Island Red bantams among others, they were revived. They were not brought over to England until the 1970s, where they were mainly supported by a group of Rare Breed enthusiasts in Bedfordshire. Despite being very pretty little birds, they remain very rare here.

Spangled Orloff, large male. Photo: John Tarren

Description

Orloffs are large, powerful looking birds, an impression accentuated by their head characteristics. German Standards require large males to weigh 3.5kg (7½lb), and the females 2.5kg (5½lb). Current Britsh Standards give very similar weights, although some older references said that some Orloff cocks were 5kg (11lb), and the best hens 4kg (8½lb). Orloff Bantams weigh about 1,000g (35oz) more or less according to age and sex.

They are quite tall, with long neck, legs and a sloping back, not surprising considering there is another Iranian breed, called the Lary that is very Malayoid, and nearby countries, such as Afghanistan and Pakistan also have many large Asil, Malay and similar breeds. The tails of Orloffs should be held just above the horizontal, and are medium length or less.

Head character is of prime importance, carrying 35/100 in the British Standards Scale of Points. The beak should be short, thick and 'well hooked'. The strong skull has prominent brow bones. Combs are of walnut or 'raspberry' type, with small hair-like feathers growing out of the surface. Wattles and lobes are small and covered by a well-developed, tri-lobed beard.

The three main plumage colour varieties are Mahogany, Spangled and White. Rarer colours are Black and Cuckoo. Details of colour are not considered important when Orloffs are being judged. Mahoganies are roughly similar to Rhode Island Reds, which has tempted some breeders over the last century to cross Mahogany Orloffs with Rhodes. This may have been a useful ploy when the Orloffs have suffered from acute inbreeding, but should be treated as a last resort. Such crosses are unlikely to have the desired head character. Spangled males are roughly as Spangled OEG, but the females are more like Speckled Sussex or Millefleurs. There is some variation in male neck hackle colour, some having a rich orange neck ground colour, others having a mahogany ground colour, matching the normal colour of all Spangled mates' back and upper wings.

Spangled Orloff, bantam female. Photo: John Tarren

Long-Tailed and Game-Related Breeds

Spangled Orloff, bantam female. Photo: John Tarren

PHOENIX (SINGLE-COMBED YOKOHAMA IN THE UK)

These are the long-tailed fowls that have been bred and exhibited in Europe since the late nineteenth century. Very few birds were transported to the West, and the breeders did not know anything, apart from a few 'travellers tales', about them. The imported birds probably included all of the Japanese breeds in this chapter, plus the Kukokashiwa and Tôtenkô in the Long-crowing Breeds chapter. After such a long sea journey, many imported birds were in less than perfect health, resulting in some last resort crossing with whichever European breeds with a reasonably impressive tail (for example, Oxford type OEG) seemed most promising. A few real O-Naga-Dori came over, and had a successful run at the shows while they lived, although the European fanciers either did not know how the Japanese experts kept them, or if they did know, had neither the time nor inclination to copy Japanese techniques. Most of the time Europeans thought themselves doing very well if they obtained tail feathers 1m (3ft) long, much less than the amazing tails seen in Japan.

Most European Phoenix were not much, if any, different from not especially good O-Naga-Don, but there is one key reason for describing them separately, Phoenix Bantams (*see below*), which have never existed in Japan, so are a purely European creation.

A French missionary, M.M. Girard, brought the first recorded Japanese long-tailed fowls to Europe in 1864. Some of these (or birds bred in France from them) were taken to Dresden in Germany for Herr Prosche. They were named Yokohamas then, because they were shipped out from the port of Yokohama.

The 1888 edition of Lewis Wright's *Book of Poultry* gives more detail of imports about 1878, mentioning Herr Hugo du Roi (Braunschweig/Brunswick), Herr Wichmann (Hamburg), Baroness Ulm-Erbach, M. Pierre de Roo (Paris) and Messrs Fowler (England).

Because of the translation difficulties from Japanese to English, French or German, plus the fact that the deals would have been done by seamen or other go-betweens, the European fanciers had to define these birds as best they could. All were lumped together as 'Yokohamas' in the UK, with German fanciers defining single-combed birds in Gold Partridge or

Black-Red Pea-Combed Phoenix, large male. Photo: John Tarren

Duckwing colour varieties as Phoenix and pea or walnut-combed birds in White or Red Saddled White as Yokohamas. They were sometimes unclear how to define single-combed Whites or pea-combed Duckwings (for example). The April 1892 'International Show' in Paris included six male and six female Yokohamas/Phoenix in combined classes, the earliest entry numbers found by the author. There were also two classes (male and female) for 'Yokohama or Japanese Longtails' at the 1904 'International Show', Alexandra Palace, London. These classes included a 'fine lot' of Duckwings shown by Rev. C.H. Hildebrand. A British Yokohama Club formed about this time, the 1904 event above may have been their first show that had a successful run-up to the outbreak of war in 1914. Its Secretary was F.J.S. Chatterton (34 Elm Park Road, Finchley, London), also a well-known poultry author and artist of the time; who tried to keep it going until 1927. In reality, it had been moribund since everyone had realized that the war would not be over by Christmas (of 1914, as optimistically imagined by many). The highest show entry figures found for the Yokohama Club were their 16–17 June 1910, Summer Show at Shrewsbury where fifty large and fourteen bantams appeared; and the Yokohama section of the 1911 Crystal Palace Show where there were thirty-five large and thirty bantams. From the effective demise of the Yokohama Club until the Rare Poultry Society was formed in 1969 the few fanciers of the breed had to enter them in the AOV classes, although a full search of surviving show records might reveal a few separate classes occasionally. Most, or possibly all, of the stock in the UK since 1970 is from Germany or The Netherlands.

There has been a breed club for Phoenix continually (since the 1950s?) in Germany, and since about 1980 they have been attempting to define a clear difference between imported Japanese O-Naga-Dori (that have arrived in greater numbers in modern times) and large and bantam 'Phönix' (German spelling) that are being redefined as a German breed. This is a long overdue clarification of the cross-breeding that has happened since the 1870s. Rolf Ismer and Hermann Falk (ref. Julia Keeling, *RPS Newsletter*, 2000) led this process. Cock tail feathers of 1m (3ft) are enough to be stylish, yet manageable for a hobbyist with a day job.

Phoenix are now bred with fairly long legs, giving an elegant, pheasant-like, style. O-Naga-Dori have shorter legs, but then they are seldom allowed to do much strutting. O-Naga-Dori have been allowed a variety of shank/foot colours, and not much attention was given to their lobes, or anything else except tail and saddle feather length. German Phönix are now required to have slate-blue shanks/feet and pure white lobes. At the time of writing (2005) British Standards had not yet been revised, but these changes were in process.

Phoenix Bantam (Single-Combed Yokohama Bantams in the UK)

Louis Neubert of Niederboritzsch, Sachsen (Saxony) is believed to have been the first to make a strain of Yokohama/Phoenix Bantams, starting about 1878. This was right at the start of their history in Europe. He crossed some smaller examples of imported Japanese long-tailed fowls (exact type unknown) with available European Black-Red/Partridge bantams. They may have been OEG (pit type) Bantams or the precursors of other broadly similar breeds such as Belgische Kriel, Dutch or Pictaves. They were first exhibited at Dresden (German National?) Show in 1898. No illustrations of his strain were available to the author, and artist-drawn pictures are not reliable. The artists often drew the ideal, not the reality.

Photographs are available of Yokohama Bantams in the UK from about 1910 onwards, and very few of them would be considered adequate now. Tail and saddle feathers seldom touched the ground on the males, and tail carriage on both sexes was too high. Really good Phoenix/Yokohama Bantams have, with few exceptions, only been seen since the 1960s.

Long-Tailed and Game-Related Breeds

*Black-Red Single-Combed Phoenix, bantam male.
Photo: John Tarren*

Black-Red/Patridge Phoenix, bantam trio. Photo: John Tarren

Description

They are stylish and pheasant-like in appearance that, when standing normally, have a slightly sloping back-line, not much above the horizontal. Tails should sweep out in a straight line from the back, gradually falling to the ground. The longest tail feathers on large Phoenix are (on good specimens) about 1m (3ft) long, perhaps 60 cm (2ft) on bantam males, the longest tail feathers on large hens perhaps 60cm (2ft), bantam hens 40cm (16in). Tips of female tail feathers usually curve down. Large weights should range from 1.5kg (3¼lb) pullets to 2.5kg (5½lb) cocks. Bantam females 700g (24oz), males 800g (28oz). There is seldom enough size difference between large and bantams.

Combs are small, single and upright with rather narrow points. Lobes are small, white and oval. Legs are described as being of 'medium' length, which is debatable. It is probably true to say that longer legs make large fowl look larger and shorter legs make bantams look smaller. Slightly shorter legs do have the advantage at a show that the saddle and tail feathers have less distance to reach the ground for that 'dragging along the ground' effect.

Special Management Requirements

High perches, well away from house walls, extra large nest boxes (preferably with a smooth interior) and frequent cleaning out are essential. Houses should be large enough for the birds to be kept inside when outside runs are wet. These outside runs should be as near to a well-kept lawn as possible. For all these conditions to be possible, it is likely that the most successful breeders will be organized and single-minded people who do not keep many other breeds, if any.

SATSUMADORI

Not all breeds fit into one of the normal categories in which types of poultry are divided, and Satsumadori could be classed with the Long-Tails or Game (fighting) breeds and their relations. With their dramatic fanned tails, they are certain to attract more fanciers who want something really exotic to strut around their garden. They are a recent arrival in Europe, first coming over circa 1980–84 to Belgium and The Netherlands. Willy Coppens, of Aalst, Belgium, was one of the leading fanciers and importers of rare breeds from Japan.

Silver Duckwing Phoenix, bantam male. Photo: John Tarren

Silver Duckwing Phoenix, bantam female. Photo: John Tarren

Satsuma was the name of the present Kagoshima and part of the Miyazaki Prefectures on the Japanese island of Kyushu during the Tokugawa Period (1603–1867). They were developed from crosses between Shamo (Japanese Game) and Shokoku (*see below*), getting their impressive build and fanned tail structure from the Shamo, and large tail and long saddle feathers from the Shokoku. The name Satsumadori was fixed for the breed about 1930, before which they were variously named O-Jidori, Kentukedori and Satsuma-Jidori.

Description

They are large, powerful and active birds. Weights range from 2.5kg (5½lb) pullets to 4kg (9lb) adult cocks. Although both sexes have large fanned tails, and males have luxurious neck and saddle hackles, the body plumage is tight-fitting. This demonstrates their dual character; they look like a long-tail, but handle like an Oriental Game breed.

Head features are similar to Shamo (pea comb, minimal wattles, light eye, and strong beak), but not as extreme. Lobes are red and shanks/feet yellow or yellow with dark shading on dark plumage colour varieties.

Colour varieties are: Black, White, Silver Duckwing and two shade versions of Black-Red, most males having a bright orange-red top colour, the remainder being a deep, rich red. The females to go with Duckwings and both shades of Black-Red males are generally brownish (a different shade for each male colour) with black neck striping, tail and back/wing lacing.

No Satsumadori Bantams have ever existed as such, but there is a Japanese bantam breed that is close, the Kawachi Yakko. At the time of writing (2005) none have been brought to Europe, America or anywhere else as far as the author was aware.

SHÔKOKU

Shôkoku are the most ancient of the Japanese long-tailed breeds. Their early history was presented by H. Oana (Japan) to the Tenth World Poultry Congress at Edinburgh, August 1954. He quoted from Kuroda Suizan who wrote about them in 1843. The birds, in their original form, were found by the Japanese Emperor's diplomatic representative to the Chinese Chusan Islands (which do not appear in the author's atlas. These and other place names have probably changed since 1843). Ningbo was also mentioned, which is south of Shanghai, so the Chusan Islands may be the islands of the coast nearby. This all happened early in the Heian Era (794–1186). The birds were then mainly used in China for cockfighting, but a similar-looking long-crowing breed, forerunners of Totenko described in the Long-Crowing Breeds chapter, was also said to exist in the same part of China at the same time. The long-crower was said to have mainly been bred around Nanking (Nanjing) and Soochow, north-west of Shanghai. The name Shôkoku was said to be the Japanese pronunciation of the Chinese district of Chang-kuo. It was fixed as the name of the poultry breed in 1206 that makes many other 'old poultry breeds', many of which were created in the late nineteenth and early twentieth centuries, seem practically new.

Silver Duckwing Shokou, male. The property of Mr Tanaka, Japan. Photo: Julia Keeling

Description

They are similar to single-combed Yokohamas/Phoenix in general appearance, indeed Shôkoku imported in the 1870s were used to make these European versions of Japanese long-tails. Because they are so similar, it is unlikely that they will attract much interest in Europe as a distinct breed now. Males' tail feathers are about 60–90cm (24–36in) long. One difference from the Phoenix is ear lobe colour, white on Phoenix, red on Shôkoku. Weights range from 1.3kg (2¾lb) pullets to 2kg (4½lb) adult cocks. There are three colour varieties that in order of popularity (most popular first) are Silver Duckwing, White and 'Five Coloured', the same as Koeyoshi in the Long-Crowing Breeds chapter.

SUMATRA

Mr. J.A.C. Butters of Roxbury, Massachusetts imported a trio of these elegant birds from Angers Point, Sumatra in 1847. In their homeland they were kept for cockfighting but, apart from the first few years, Sumatras have been regarded as an ornamental breed, be it for the show hall or just to strut around the country garden in America and Europe. Sumatra hens are excellent sitters and mothers, a trait that has helped prevent them from going extinct. They were once called 'Sumatra Game', but the 'Game' part is usually dropped now. Sumatras are very much an ornamental 'long-tail' in the minds of poultry exhibitors.

Sumatras were admitted to American Standards in 1883, but available records suggest that they did not attract much interest. The *American Poultry Journal Yearbook* for 1913 did not include a specialist breed club for them, nor were any Sumatras mentioned in the results given for major American shows in the 1912–13 winter season. Early imports were more variable in colour, especially the males, many of which had red or orange in neck and saddle feathering. As they were refined by exhibition breeders, only Blacks with a very prominent green sheen became acceptable. There is (or was circa 2000 anyway) the American Sumatra Association now, but the author has not discovered when it was founded.

Mr Frederick R. Heaton of Norwich imported the first Sumatras to England from Toronto in 1901 or 1902. A 'Sumatra Game Club' was formed that put on a few impressive displays at their Annual Shows at Crystal Palace in the years leading up to the First World War. They had four classes each time (cock, hen, cockerel, pullet) with totals of forty-six entries in 1910, forty-four in 1912 but down to twenty-two in 1913. Because of

Sumatra, large male. Photo: John Tarren

Sumatra, large female. Photo: John Tarren

the hard economic conditions of the 1920s and 1930s the number of Sumatra breeders declined, although the club struggled on until 1939.

Photographs of winning Sumatras before 1911, show birds with very poor tail or saddle feathers compared to the best of today's birds. Dr Royden of Norfolk, for many years Sumatra Club Secretary, recalled (1935 *FW Yearbook*) the 1912 Crystal Palace Show, when significant progress was made:

> There was quite a disturbance in the club regarding the length of saddle hackles, there were several cockerels shown with saddle hackles of about five inches in length. Some of us said they had been crossed [with Yokohamas?] and were not true-bred Sumatras. Mr. Eaton and the Rev. Serjeantson defended the long feathers saying they had produced them by selection.

After further discussion, including a suspension of the meeting to inspect the birds in the show, the club members agreed to alter the standard to read: 'Saddle hackles – the longer the better, provided they are broad.' The proviso was presumably to avoid people crossing them with Yokohamas that have long narrow saddle feathers.

Mr Eaton's original strain was kept going into the 1970s by Mr John Spedan Lewis, well known in the UK as the founder of the 'John Lewis Partnership' retail chain. His main hobby at his country estate (Leckford, Hampshire) was the breeding of waterfowl, and the Sumatras were kept for hatching them. Some were passed on to another waterfowl breeder, Mr Jack Williams. Either of these breeders may have crossed them with Silkies; as when other fanciers started to breed this strain as show birds (rather than just broodies) they noticed that they had excellent dark faces, but not much pheasant style. During the lean years, other breeders used Black OEG, sometimes necessary when pure Sumatras were difficult to find. This produced 'Sumatras' with red eyes, red faces and short tails. Since the 1980s, Sumatras have generally been of better quality than they ever had been before.

Sumatra Bantam

They seem to have been made independently in Germany, The Netherlands, the USA and the UK several times since the first attempts in the 1920s. It is believed that all current strains are post-war, probably made since 1970.

All attempts have naturally included undersized large Sumatras that have been crossed with a selection of suitable bantams including: Black OEG, Birchen and Black Modern Game, Black-Red Phoenix/Yokohamas, Black

Sumatra, bantam male. Photo: John Tarren

Sumatra, bantam female. Photo: John Tarren

Rosecombs, Black German Bantams (Deutsche Zwerghuhner), Rhode Island Reds and Black Hamburgs (the last in Germany).

Since the 1990s (in the UK at least), Sumatra Bantams have been superior to the large fowl for elegance and style. Many show organizers provide extra large cages to allow space for their tails to be displayed without being damaged, but their body size is small.

Description

General style is pheasant like, with a long tail that extends out in a straight horizontal line from the back and then sweeps down to the ground. The females' two longest tail feathers turn down at the end. Male saddle hackles touch the ground, even dragging along slightly on some of the best specimens. This is much longer than they were before 1914.

Apart from a strong beak, Sumatras have refined heads. They ideally have small, neat pea combs. Some males have rather high, irregularly shaped combs, one of the most common faults in the breed. Lobes and wattles are as small as possible. Beak, eyes and skin should all be as dark-coloured as possible. Large fowl weights range from 1.8kg (4lb) pullets to 2.7kg (6lb) adult cocks with bantams ranging from 625g (22oz) to 735g (26oz).

Black was for a long time the only standard colour, and it should be an intense green-black. Blues have been bred since the 1980s, and are usually a dark shade. White Sumatras are even more recent. It is particularly important to retain the dark skin, eye and shank colour on Blues and Whites to avoid any tendency to confusion with Yokohamas or Phoenix.

Special Management Requirements

As may be imagined these mainly involve housing and management to preserve tail condition, such as high perches. Their black or blue plumage can be nutritionally demanding, so a high protein diet, perhaps starting with turkey starter crumbs might be advisable.

TWENTSE HOEN/ KRAIENKÖPPE

This breed was developed on both sides of the border between The Netherlands and Germany. Twentse is the Dutch name and Kraienköppe the German name for the same breed. As if having two names for one breed was not enough cause for confusion, the breed that is called the Breda in most of the world is named 'Kraaikoppen' in The Netherlands.

Gold Twentse/Kraienköppe, large male. Photo: John Tarren

Gold Twentse/Kraienköppe, large female. Photo: John Tarren

Both versions of 'the K word' mean crow-headed, a reference to both breeds having minimal combs, thus producing the alleged resemblance to crows. It would help if each breed was called Breda and Twentse universally.

They originated in the Twentse area round Enschede in The Netherlands, and over the border in the German Grafshaft (roughly equivalent to a UK county) of Bentheim. Like many other breeds, they started from crosses between previously existing local chickens, in this case local to Twentse and Bentheim, with Asiatic imports from the mid-nineteenth century. It took a long time from the first crosses (circa 1855?) until the first 'Twentse Grijzen' (Twente Greys) were exhibited at the 1892 Enschede Show. In the earlier part of this period, the Asiatics imported would have been Malay types, some, but not all, of which would have been reliable fighters. Cockfighting was very popular on both sides of the Belgian/French border around Lille, and these Orientals were crossed with Combattant du Nord Game to increase size. Any 'duds' would have been passed on to the Twente or Bentheim areas for table bird crossing.

As might be guessed from the name, Twentse Grijzen, the first birds shown by the Lazonder Brothers were silver hackled blacks, a popular colour in Belgian Game. It is not known if the females were coloured as Silver Duckwing Game, Grey Game or indeterminate variations in between. The Silver Duckwing pattern and large fanned tails seen today were fixed in the 1920s when Silver Duckwing Leghorns were added to the mix; primarily to improve egg production. In Germany, they were first shown at Hanover in 1925 by J. Wieking. For a few years around 1930 there were several more colour varieties: Barred, Black, Wheaten and White. Never very popular anyway, they all rapidly died out. The second permanent colour, the Gold Partridge was recognized in the 1950s. Blue versions of both Golds and Silvers (Greys) are recent (circa 1990?) creations.

There were two earlier importations to

Silver Twentse/Kraienköppe, large male. Photo: John Tarren

Britain (1928 and 1930) that rapidly died out; so all the current UK stock is from importations since 1980. They have only been kept by a few enthusiasts here, notably Rare Poultry Society Committee members Brian Sands, Lincolnshire and Brian Ward, Bedfordshire.

Twentse/Kraienköppe Bantam

Attempts to make these started in the late 1930s, but were halted by the Second World War. They were made from undersized large Twentse Hoenders, Malay Bantams and OEG Bantams. It is not clear if any of the first strain survived through to the 1950s. The post-war revival was led by Mr Siemerink, Almelo, Netherlands and Bernhard Ahlbrand, Recklinghausen, Germany. They were recognized by the German Show Standards in 1957. Also in 1957, another German fancier, Ernst Erdelkamp, made the first Gold Partridge Bantams. He started with a sex-linked cross: Partridge OEG Bantam × Silver Twentse/Kraienköppe Bantam. The silver males from this mating were discarded, the gold pullets then being mated with a Black-Red Malay Bantam male. More years of selective breeding were needed until Golds were good enough to be recognized in 1963.

Some, mostly Silvers, were imported to Britain in the 1970s. There was a group of keen breeders for a few years, but interest

Silver Twentse/Kraienköppe, bantam male. Photo: John Tarren

Silver Twentse/Kraienköppe, bantam female. Photo: John Tarren

soon fizzled out. Since 2000 even major shows have only had two or three Twentse bantams at most.

Description
They are an upright, elegant and stylish breed that would look very fine strutting around anyone's country garden. This is as expected from the Game, Malay and Leghorn ancestors. Their tails, especially the males, are large, fanned and carried about 30–40 degrees above horizontal. Females' tails are not fanned as much as those of males. The breast should be well-rounded, sometimes lacking on the bantams. Large fowl weigh from 1.75kg (3¾lb) pullets up to 3kg (6½lb) cocks. Bantam males should be 850g (30oz), females 750g (26oz). Twentse males have narrow walnut combs and small wattles, both almost absent on females. Lobes are red, eyes orange and legs yellow.

All varieties are different colour versions of the same classic Black-Red/Partridge pattern. It would appear from most specimens seen here that double mating is not employed. They are nearly 'cock-breeders'. Cockerels usually have solid black breasts, sometimes a little spotted with white (Silvers or Blue-Silvers) or gold (Golds or Blue-Golds). Females seldom have really good salmon breasts (required on all varieties), the back colour usually extending down from the wings to the sides of the breast giving a dull appearance.

YOKOHAMA (JOKOHAMA IN GERMANY)

The walnut-combed Red Saddled White Yokohama is not a recognized breed in Japan, although they came from there, possibly with some later crossing with European breeds. All long-tailed fowls were named 'Yokohamas' in the UK, but German breeders (in an attempt to make some sense of the variety of imports arriving in the 1870s) named the single-combed birds, mostly with white lobes and very long tails, 'Phönix' and the pea or walnut-combed birds with red lobes and not quite as long tails, 'Jokohamas'. All the early imports had been sent from the major city and port of Yokohama, hence the name.

As described earlier in this chapter, Phoenix/Phönix were predominantly O-Naga-Dori, but with some crossing with European breeds (OEG and others?) to overcome the rarity, inbreeding and health problems the European fanciers had to cope with at the time. Yoko-

Long-Tailed and Game-Related Breeds

Yokohama, idealized pairs. Artist: Kurt Zander

hamas/Jokohamas may have been crosses involving Minohiki, Satsumadori and who knows what else? The main European breeder associated with the Red Saddled White birds that in the early days came with either walnut or pea combs, was Hugo du Roi of Braunschweig (Brunswick), the first President of the Bundes Deutscher Geflügelzüchter (German National Poultry Club). As Red-Saddled White Yokohamas are not a recognized breed in Japan there has been some speculation as to whether Hugo du Roi made them (by accident?) in the course of the crossing that all the first importers were resorting to (to make the best of the few birds available), or did Red Saddles arrive as such. Wilfred Detering's book *Kämpfer und Zwerg-Kämpfer der Welt* (1983) includes a reproduction of a watercolour of a pair of Red Saddles, the caption of which indicates they were direct imports: Erste Importe von Präsident Hugo du Roi, Braunschweig. Aquarell nach dem Leben von Gustav Mützel, dem Illustrator von 'Brehms Tierleben', Berlin 1882.

Just because Red-Saddled White Yokohamas have not been recorded as a recognized variety in Japan since regular communications at a high level between poultry experts in Japan and the Western world were possible (say from the 1920s), this does not make it impossible for such birds to have existed in Japan circa 1870. They may have been chance crossbreeds in the Japan of 1870, or may even have been a localized breed that later died out.

White long-tailed fowls have regularly appeared in both Phoenix/O-Naga-Dori and Yokohama types. This has probably caused some confusion over the years, although in recent years the different types, ear lobe and shank/feet colour and comb type have been fixed to clarify matters.

Harrison Weir (*The Poultry Book*, 1902 and 1905) wrote, under the heading 'White Yokohama Fowl': 'In some parts of Saxony and Germany they are raised to great perfection, and clubs have been instituted for their production and possible improvement.' Saxony, along with many other German states were more or less independent before they were unified into a single nation state by Otto von Bismarck, Chancellor of Prussia, 1862–71.

The genetics of the Red-Saddled White plumage pattern in not fully explained in any of the usual books on poultry genetics. Genes that may be present to give this combination of colour and pattern may include: Gold (sex-linked), Dominant white, Spangling, Mahogany and there obviously must be one gene on the 'E locus', possibly eb 'brown'.

Red-Saddled White Yokohamas do not seem to have been bred in the UK until recent times, which is surprising for such a beautiful breed that would have been readily available in Germany. There may have been a few here circa 1905–10, possibly imported by Mrs Prideaux a well-known breeder before the First World War. No more seem to have been recorded at British shows until the early 1970s. From about 1964, groups of British fanciers started to visit the large shows in Germany and The Netherlands, among which were Rex Woods, President of the PCGB, 1967 and 1969–71. He brought hatching eggs over from Germany and (later) Switzerland. The author is not certain exactly when this happened, but Rex was advertising Red Saddled White Yokohamas for sale in the 1968 PCGB Yearbook. Since then, they have remained a 'Rare Breed', but have probably been kept in greater numbers as ornamental birds for country gardens than the numbers seen at the shows would suggest. Although not as much trouble to keep as the very long-tailed O-Naga-Dori, they are still one of the more difficult breeds to manage and prepare for exhibition.

Yokohama Bantam

As large Red-Saddled White Yokohamas are (or the best of them anyway) one of the most beautiful of all fancy poultry breeds, it is surprising to learn that the bantam version has had such a chequered history.

Hans Sack of Leipzig-Connewitz made the

Long-Tailed and Game-Related Breeds

first attempt in the 1920s. His programme started from crosses of undersized large Red Saddles and White Deutsche Zwerghühner (German Bantams). Many, probably most, of the large Red Saddles were undersized. This may be a reason for the bantams not becoming very popular – there was not enough difference in size between 'large fowl' and 'bantams'. Herr Sack first exhibited his new creation at the third German 'Nationalen Zwerghuhnshau' (National Bantam Show) at Essen. This strain certainly did not survive the Second World War, and may not even have lasted that long, such were the hard times of hyper-inflation between the wars.

After the war, another German fancier, Amo Beyrich, tried again (circa 1956), using the same cross as above, plus Pile Modern Game Bantams. The author does not know how long this strain lasted. Yet another German fancier, Werner Urbanski, did much the same process in the mid-1960s. They were

Red Saddled White Yokohama, large male. Photo: John Tarren

White Yokohama, large male. Photo: John Tarren

White Yokohama, large female. Photo: John Tarren

standardized in West Germany in 1968. East German fanciers also had them. It is not known if they were made independently. Although general history makes much of the Berlin Wall and the complete separation of East and West German societies, the poultry fanciers of both parts of Germany seem to have been in regular contact. Small livestock showing seems to have been an approved hobby in the DDR, so perhaps special arrangements were made for fanciers.

A small group of fanciers in Germany and The Netherlands were working on yet another remake of Red-Saddled Yokohamas (and Sumatra Bantams, simultaneously) about 1978–84, and it is these that are the current strain at the time of writing, 2005. They have been bred in the UK since the 1990s, the main breeder and exhibitor here being Richard Billson of Great Glen, near Leicester, and Rare Poultry Society Secretary. Richard has bred and sold many trios of beautiful (but not quite as good as those he kept!) Red-Saddled White Yokohama Bantams, but very few of the buyers have exhibited them.

Description

Yokohamas differ from Phoenix and O-Naga-Dori in (ideally) being larger, taller and more 'gamey'. Many are slim to the point of puniness, so constant selection for size and fitness is essential. Some idealized illustrations show great length of tail, but the reality is more limited. Saddle and tail feather reach the ground and drag along slightly, but not much more than that.

Pea and walnut combs are both permitted on White and Red-Saddled White Yokohamas, but only walnut combs are usually seen on the latter. Wattles are small or even absent. They have a tendency towards Oriental Game head character, but are not as powerfully built.

On both sexes, the plumage of neck, legs and tail of both sexes is pure white. The 'red' parts of males are a deeper, more mahogany shade, than the softer colour on females. The breast of both sexes is red with white 'kite-shaped' tips on enough of the feathers to give an attractive appearance. Wing feathering (apart from flights that are pure white) on females is the same as the breast, but a darker shade of red. Female back plumage is usually clear white, but some red may spread up. The shoulders and upper wings of males are deep, rich glossy mahogany. Yokohamas have red lobes, yellow beak, shanks and feet.

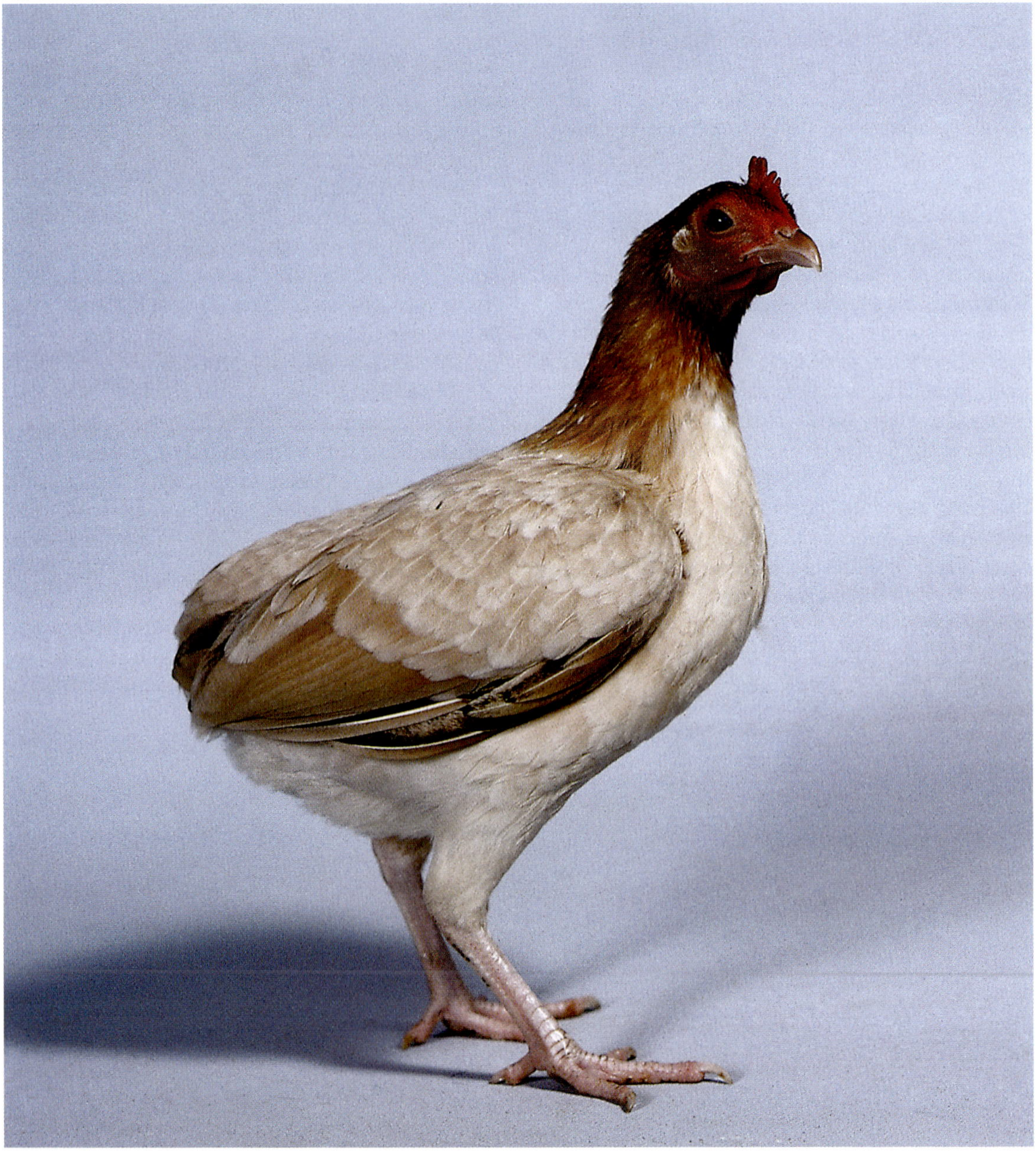

CHAPTER 4

Rumpless Breeds

There are a number of these from around the world, each more or less related to normally tailed local breeds.

This chapter describes all recognized rumpless breeds except Rumpless Araucanas. Famed for their blue-shelled eggs, Araucanas, and the related (US) Ameraucanas, are popular among hobbyists, and are well-documented elsewhere.

The breeds covered in this chapter are:

Earlier types of Rumpless
Ardennaise Sans-Queue/Ardenner Bolstaart/Rumpless Ardennes – Belgium
Barbu d'Everberg/Everbergse Baardkriel – Belgium
Barbu du Grubbe/Grubbese Baardkriel – Belgium
Drentse Bolstaart/Rumpless Drente – Netherlands
Kaulhuhn and Zwerg-Kaulhuhn – Germany
Ruhlaer Zwerg-Kaulhuhn – Germany
Rumpless Game, Modern Game and OEG types – UK and (independent!) Isle of Man
Uzara-O – Japan.

EARLIER TYPES OF RUMPLESS

Ulisse Aldrovandi described and illustrated rumpless *Persian Fowl* in his book back in the year 1600 (modern English language edition, *Aldrovandi on Chickens*, L.R. Lind, 1963). The pair of birds featured need not be fully described here as they were no doubt just rumpless versions of randomly bred farmyard chickens. All we can say now, is that this proves that such birds were known in Europe (Aldrovandi was based at the University of Bologna) then.

The first major illustrated poultry book of the Victorian period, *Wingfield and Johnson's* (1853) gave two (and a bit) pages, including an illustration, to 'Rumpless Fowl'. These learned gentlemen cited two places where 'Rumpkins' had been observed, plus a mention of Aldrovandi's *Persian Fowls*. The Rev. J. Clayton had noticed some in Virginia in 1693. It was not certain then where they had come from – some Virginian locals said England (or Isle of Man?), but this was doubted. In the light of knowledge gained over the intervening 150+ years, they might have come from somewhere in the Caribbean or South America, that is, one of the Araucana ancestral stocks. Ceylonese (now Sri Lankan) 'Wallikikilli are also mentioned. Although some rumpless fowls were certainly found in Sri Lanka, a Mr. E.L. Layard (who lived there in 1850) was sure they were a recent introduction. He thought they had been brought in from further east. This would suggest a connection with the Japanese Uzara-O. Wingfield and Johnson concluded by mentioning two current (in 1853) breeders of Rumpless, Mr Beetenson of Birmingham and Mr Nolan of Dublin. These birds were typical light-breed size 2.5kg (5½lb) cocks and hens 'rather under 5lbs (2.2kg)'. They had bluish-white

107

legs, rose or single combs, occasionally with small crests, and appeared in several colour varieties: Black, Black-Red/Partridge and White with irregular amounts of spangling or pencilling. These birds could have been related to one or more of the Rumpless Ardenner, Drentse or Kaulhuhn detailed in this chapter.

Earlier editions of *Wright's Book of Poultry* repeat much of the above information, plus a few more details of breeders from about 1870. Mr Henry Higgs of Lewes, Sussex and Miss Bush of Clifton (Bristol? or one of the other Cliftons?) were the breeders, and their birds were given the name 'Gondooks'. They were fairly small, but evidently remarkable fowls. Imagine, if you will, a Black Rumpless Sultan or Burmese: five-toed, feather-legged, vulture-hocked, crested and bearded. This is perhaps an extinct fancy breed that would be worth remaking. Wright also mentioned a 'West Indian gentleman' he met at the 1872 Crystal Palace Show who informed him that many fowls in his homeland (island not specified) were rumpless.

Harrison Weir (UK 1902, US 1905) briefly repeated the earliest sources and added that rumpless fowls of various shapes and sizes were known in Belgium (*see below*) and parts of Africa. There are brief, and none too clear, references to rumpless fowls in America. Although 'Rumpless' are included in American Bantam Standards, they are not in the APA Standard of Perfection, and scarcly exist. The ABA Standard form appears to be broadly similar to bantam Ardennes/Drentse/Kaulhuhn.

W.F. Entwisle (*Bantams*, 1894) went into more detail on Rumpless Bantams and the then 'experimental' matings tried by W.B. Tegetmeier for Charles Darwin's world-changing books. Tegetmeier crossed a (large) White Poland cock with a small White Rumpless hen that he obtained from Turkey. After three years of selective breeding he had made a strain of large White Rumpless Polands. These, like the Gondooks described above, also sound interesting enough to make again.

Mr Tegetmeier also used his original Turkish White Rumpless to produce Rumpless Nankin Bantams, obviously by mating with a normal-tailed Nankin cock. Entwisle went on to mention the range of shapes and sizes of Rumpless Bantams known in Britain in the 1880s, including Rumpless Game and Rumpless Booted (as Ruhlaer Zwerg-Kaulhuhn *see* p.111). A Mr Garnet produced (approximate) Rumpless Silver-Laced Sebrights in 1889 that are sadly long extinct.

ARDENNAISES SANS-QUEUE/ ARDENNER BOLSTAART/ RUMPLESS ARDENNES

They are obviously rumpless versions of the tailed large and bantam Ardennes, and are closely related to Dutch Drentse Bolstaart and German Kaulhuhner. They, or something very much like them, have cropped up for centuries (as seen above), but have only existed as a fixed breed since the time of the First World War. Georges Lamarche and Dr Van Bogaert are mentioned as early breeders of the standard type as large fowl and Théo de Lame with the bantam version. They are the same as normal Ardennes (*see* European Light Breeds chapter), but minus the tail.

BARBU D'EVERBERG/ EVERBERGSE BAARDKRIEL

These are Rumpless Barbu d'Uccles/Ukkelse Baardkriel in all but name, and are also practically identical to the bearded version of Ruhlaer Zwerg-Kaulhuhn. They were first made by Robert Pauwels at Everberg Castle near Brussels in about 1906. As might be guessed, they were made from crosses between Barbu d'Uccles and (details unknown) rumpless chickens. First shown in 1918, progress was hindered from 1914 until 1945 because of world events. Georges Lamarche is credited being the main breeder who revived Barbu d'Everberg in the late 1940s and early 1950s. Not many other people have shared Robert

Rumpless Breeds

Pauwels' and Georges Lamarche's enthusiasm for them, as Barbu d'Everberg have only just managed to survive through to 2005. In the UK they are not officially 'Rare' because they are covered by the British Belgian Bantam Club, not the Rare Poultry Society.

Body shape and plumage colour varieties are as Barbu d'Uccles and the smaller strains of Booted except for the tail. Although nearly the same as Bearded Ruhlaer Zwerg-Kaulhuhner, Everbergs are a touch more compact.

BARBU DU GRUBBE/ GRUBBESE BAARDKRIEL

Equivalent to rumpless Barbu d'Anvers/ Antwerpse Baardkriel, Grubbes were also made by Robert Pauwels about 1904 from normal tailed Barbu d'Anvers and some unknown rumpless bantams. Some Black 'Barbu d'Grubbe' were among the display of Belgian Bantams entered in the 1911 Crystal Palace Show. The Barbu d'Anvers and Barbu d'Uccles at the same show quickly attracted a number of British breeders, but the rumpless Barbu du Grubbe never caught on. However, they survived back in Belgium, and a new generation of British fanciers brought some over in the 1980s. As with the Barbu d'Everberg, Barbu du Grubbe are covered by the British Belgian Bantam Club, so are not officially 'Rare' despite their low UK numbers.

Barbu du Grubbe have the same rose comb, beard, small size and jaunty carriage as Barbu d'Anvers; and they are permitted in the same colour varieties.

DRENTSE BOLSTAART/ RUMPLESS DRENTE FOWL

These are the Dutch equivalent of Belgian Ardenner Bolstaart and German Kaulhühner.

Cuckoo Barbu du Grubbe, male. Photo: John Tarren

White Barbu du Grubbe, male. Photo: John Tarren

White Barbu du Grubbe, female. Photo: John Tarren

Blue Barbu du Grubbe, female. Photo: Hans Schippers

All details, except for the tail, are as for normal-tailed Drentse described in the European Light Breeds chapter. Drentse Bolstaarts and Kaulhühner both have white ear lobes, giving at least one difference from the red-lobed Ardenner Bolstaarts. There are large and bantam versions.

KAULHUHN AND ZWERG-KAULHUHN

Closely related to Ardenner Bolstaart and Drentse Bolstaart, Kaulhühner were refined into a standard breed in the 1890s, the main centres being Sachsen (Saxony) and Thüringen. Two of the pioneer breeders mentioned were U. Reber and K. Huth. The bantam version was made around the same time, L. Aßmus of Frankfurt/Main being the leading early breeder.

Kaulhühner are a typical light breed with modest single combs, white ear lobes and slate coloured shanks and feet. There are a lot of colour varieties, including Black, Blue, Lavender, White, Barred, Black-Mottled and all the colour versions of the Black-Red pattern, in this case with Asiatic-type pencilled

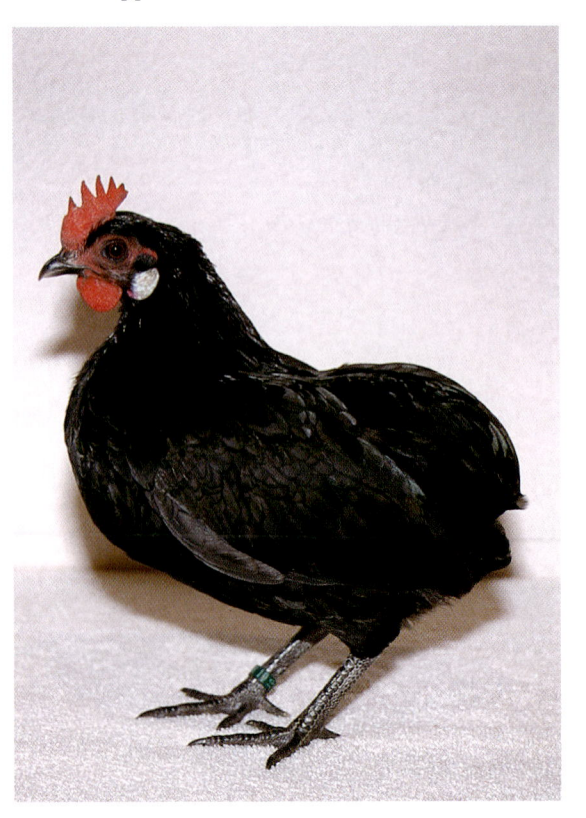

Black Drentse Bolstaart / Rumpleness Drente, female. Photo: Hans Schippers

females. Various colour combinations of the Spangled pattern are also known, probably an indication of part Thüringian Bearded ancestry.

RUHLAER ZWERG-KAULHUHN (GERMANY) OR RUMPLESS BOOTED (UK)

Equivalent to Rumpless Federfüßige Zwerghühner/Booted, they are a standard breed in Germany but have been recorded elswhere. Mr Frank Rice of (the hamlet) Stanstead, near (the slightly larger village) of Glemsford, Suffolk had 'Rumpless Booted' in 1901. He exhibited a (Black?) Mottled hen at the 1901 Liverpool Show, and as he also bred Black and White normal Booteds, he may have had the rumpless version in these colours as well.

Wandelt and Wolters (1998) give 1931 as their first appearance in Germany, these having been made from Booteds, Zwerg-Kaulhühner and Thüringer Zwerg-Barthühner. It is not known if German breeders also had stock from Belgium, the UK or elsewhere.

They are the same size, shape and colours as normal Booteds, and are permitted in both bearded and non-bearded forms. There would obviously be very little, if any, difference between a bearded Ruhlaer Zwerg-Kaulhuhn and a Barbu d'Everberg.

RUMPLESS GAME (MODERN AND OLD ENGLISH GAME TYPES)

These are traditionally associated with the Isle of Man that, as its residents frequently point out, is a (British) Crown possesion, but has its own parliament and is not part of the UK. As the rest of this chapter makes clear, rumpless fowls have cropped up in several places around the world, but apart from South America (Araucanas and similar), have never attracted many breeders. The only reason Rumpless Game have become a popular breed on the Isle of Man is because the tailless Manx Cat has become a symbol of the

Black Rhulaer Zwerg-Kaulhuhn / Rumpless Booted, male. Photo: John Tarren

island. According to legend Manx Cats date back at least to the sixteenth century, and they were certainly established by the midnineteenth century when regular steamship services changed the Isle of Man from a remote community into a tourist resort. Several farms opened up for cream teas in the summer tourist season, and the tourists (mostly from Liverpool?) expected to see a Manx Cat and a flock of Rumpie chickens wandering around the garden. It could be argued that the true original utility purpose of both cats and chickens were to separate tourists from their money. Edward Brown (*Pleasurable Poultry Keeping*, 1898) noted that 'at one time' Rumpless were common on the Isle of Arran, off the coast of Scotland, not far away from the Isle of Man.

Information on Rumpless Game has been difficult to find. They were not even fully described in *British Poultry Standards* until

Rumpless Breeds

the 1982 (*Bantams Only*) and 1997 (*Large and Bantam*) editions. Earlier editions just gave them a very brief paragraph in the 'Other Breeds' section. However, there is other definite evidence that at least two types had been established on the Isle of Man (and on the British mainland?) in the 1890s.

The original type seems to have died out. They were fairly, but not very, long in the neck, body and legs, and not as tight in body plumage as later types. In overall shape they seem to have been somewhere between Ardenner Bolstaart and a rumpless version of Spanish Game or small Oxford-type Old English Game. One can only guess at their weights, probably about a kilogram (2lb). The distinct Modern and Carlisle OEG types detailed below tend to revert to this original type without frequent backcrosses.

This original type was (presumably) crossed with normal-tailed Modern Game Bantams to create the next version. Mr Mundle, of Ballaugh, Isle of Man, successfully exhibited Black-Red (Modern type) Rumpless Game Bantams in the 1890s. These seem to have died out by 1914, but some fanciers on Man started to recreate them from about 2000.

Rumpless Game of both Modern and OEG type were usually dubbed (surgical removal of comb of males) by Manx-based exhibitors, but as Rumpless were not usually dubbed on the mainland, this ceased when island fanciers started to enter mainland shows regularly.

A few large Rumpless Carlisle-type Old English Game have been seen at the shows since the 1970s, most of them Black-Red/Partridge. Few of them were from the Isle of Man.

Rumpless OEG-type Bantams are currently covered by the Rare Poultry Society. The impressive displays seen at some UK

Brown-Red Rumpless Game, bantam male. Photo: John Tarren

Wheaten Rumpless Game, bantam female. Photo: John Tarren

shows over the last decade (1995–2005) might make some people wonder why they are classed as 'Rare', but all these birds come from a fairly modest number of fanciers. Thankfully, from a conservation point of view, the current breeders are well scattered, from Kent to Cornwall and well up in the north of England.

As it only takes two or three generations of selective breeding after a cross with normal-tailed OEG Bantams to start a new strain of Rumpless Game, their future should be secure, and the range of colour varieties should continue to expand. At small shows where there are no specialist Rare Breed judges officiating, Rumpless Game stand a good chance of winning the 'Best Rare Breed' award under all-round judges.

UZARA-O

These originated in the Kochi Prefecture in the south-west of Japan. They are related to Ko-Jidori (Japanese 'country bantams') and Chabo (UK 'Japanese Bantams'). Uzara-O have modest upright single combs, white ear lobes and shortish yellow shanks and feet. Standard weights are: cock 675g (24oz), cockerel and hen 600g (21oz), pullet 563g (20oz). The main colour varieties are Black-Red/Partridge, White and 'Goishi' (irregular black and white mixture, as with Exchequer Leghorns).

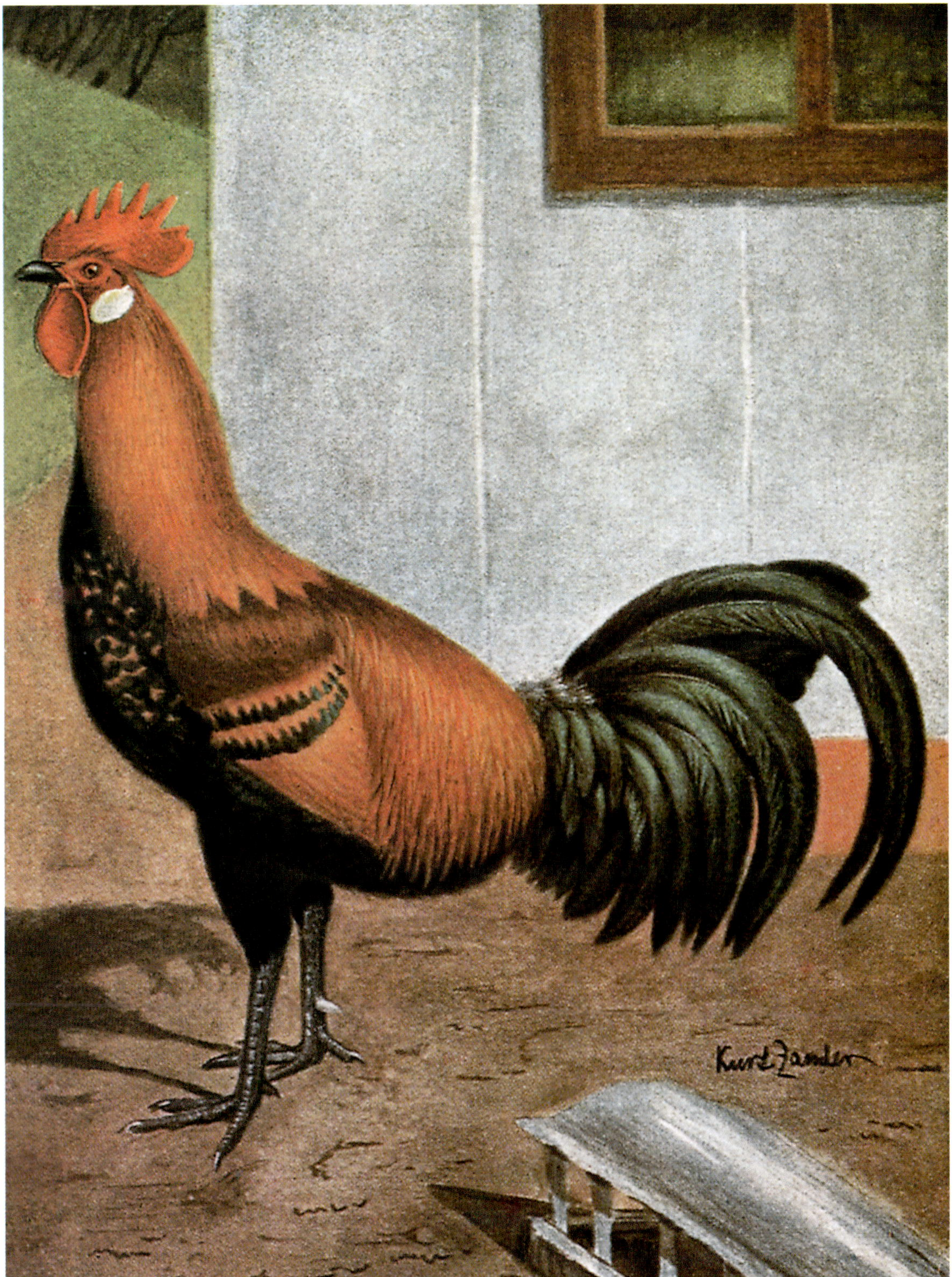
Kurt Zander

CHAPTER 5

Long-Crowing Breeds

Crowing competitions are a cultural phenomenon that seem to have developed in a number of far-flung parts of the world, apparently independently. This group of breeds may have started in south-east Asia, with the Indonesian breeds that are named below, but are not fully described because of lack of available information. There may be ancestral links between the long-crowers, the long-tailed breeds and Araucanas.

Breeds described in this chapter:

Bergischer Kräher – Germany
Denizli – Turkey
Jurlowskije Golosistye – Russia
Koeyoshi – Japan
Kurokashiwa – Japan
Tômaru – Japan
Tôtenko – Japan.

All of the above occasionally appear at British, Dutch and German shows, where the judges assess them (as well as their know-ledge permits) on their appearance rather than crowing ability. Most judges will look more favourably on any cocks that 'perform' while judging is in progress, but will probably not know the fine details of judging the crowing.

Wandelt and Wolters (*Handbuch der Hühnerrassen*, 1996) describe several more long-crowers that are listed here, but not fully described because little is known about them, and none are likely to be seen at a European show in the near future.

They are:

Ayam Pelung – Indonesia ('Ayam' means chicken in Indonesian)
Ayam Ratiah – Indonesia
Ayam Tertawa – Indonesia
Ayam Yungkilok – Indonesia
Bosnian Crower – Bosnia-Herzegovina and other Balkan States
Chan – Japan
Galo Musico – Brazil
Katai Parsi – Indonesia
Katai Sutera – Indonesia.

BERGISCHER KRÄHER

The Bergischer Land is a hilly area to the south of Düsseldorf and Dortmund. Crowing cockerels raised the alarm when Graf (Count) von Berg's castle was subject to a surprise attack in the year 1190, thus making crowing cockerels part of local legend. It is most unlikely that the Bergischer Kräher breed, as now known, existed in 1190, so we can only guess at the real medieval history of the breed.

There is an annual 'Kräherfest', probably with additional smaller-scale competitions, in the Bergischer Land, in which the length of each crow is timed. Good specimens can keep the end note going to give an overall time of 30 seconds. It is not known how long such competitions have been held, but presumably it would have been difficult to judge them with any accuracy before suitable clocks were available, possibly during the eighteenth century.

Turning from crowing competitions to conventional poultry shows, the first Bergischer Krähers known to have been exhibited were at the 1853 Görlitz Show. A breed standard was passed for them in 1885, and a breed club was formed in 1916. Although not very far from major cities, Bergische Land is a rural area which seems not to have suffered as much as other parts of Germany in the closing stages of the Second World War.

As has always been the case, there have always been more Bergische Krähers in their fairly small home region than all of the rest of the world. The few of these (and other long-crowers) seen (and heard!) at shows elsewhere generally attract a lot of amused interest.

Bergische Zwerg-Kräher/ Bergischer Crower Bantam

This is the only long-crowing breed to have a bantam version. The first person to attempt to make Bergische Zwerg-Krähem was Herr Fritz Gebigke of Lennep (in Sachsen/Saxony, not the Bergische Land in Nordrhein-Westfalen). His breeding programme took place during 1920–25, and started from a cross between undersized large Bergische Krähern and (Black?) Rosecomb Bantams. His strain looked wonderful, small and with the correct plumage pattern and colour. Unfortunately, the crossing lost the long-crowing. No one, especially not in Bergischer Land, was going to accept a miniature version of the breed if it didn't perform; long-crowing was their raison d'être.

More successful were Wilhelm Küppersbusch (Velbert) and Emil Schlesinger (Schneeberg) during the 1930s. They simply selected the smallest specimens of the large fowl, incubated small pullets' eggs, and inbred to reduce the size without losing crowing ability. This has resulted in a miniature that performs well, has the correct plumage colour, but is usually larger than most would like. They remain rare, with many major shows in Germany having none entered, or if any, only from one or two exhibitors.

Bergischer Kräher, male. Artist: Kurt Zander

Description

In general appearance, they are a tall and active looking light breed, standing well up with comparatively low tail carriage. Bergische Krähers have a curved back (like Malays), probably associated with enlarged lung and/or air-sac capacity; thereby assisting the long crow. Weights range from 2kg (5½lb) pullets up to 3.5kg (7¾lb) adult cocks for the large fowl; 800g (28oz) female, 900g (32oz) male, for the bantams.

They have medium-sized, upright, single combs; smallish, oval, white ear lobes; orange eyes and slate blue shanks/feet. The only plumage colour is 'Schwarz-goldbraungedobbelt', which is almost unique. Bergische Schlotterkamm have the silver version of the same pattern. It is mostly black, with gold showing on the wings, breast and neck hackles of both sexes and the saddles of males. The pattern is a very melanistic (extra black) version of Gold Laced.

DENIZLI

Denizli is a city in south-western Turkey, close enough to ancient main trading routes, and not too far from the coast to benefit from goods being shipped around the Mediterranean, for the Denizli to have received long-crowers from the Far East (Indonesia? Japan?), or to have passed stock further north (Juriowskije Golosistye) or north-west (Bergische Kräher) to play a part in the development of other long-crowers.

A reasonably good Denizli crow is about 20–25 seconds long, with the longest ever recorded performance being 45 seconds. Unfortunately, the author did not have any details of the numbers bred or competition rules in Turkey. They became known to fanciers in western Europe by German poultry experts Wolfgang and Brigit Vits when they toured a Turkish Government Denizli Fowl breeding establishment in the 1980s. They noticed that the city has a crowing cock on its coat of arms and the breed was well known locally.

They are a medium-sized breed, ranging from 2kg (5½lb) pullets to 3.5kg (7¾lb) adult cocks. General build is very conventional apart from having (like Bergische Kräher) a curved back. Most Denizli have single combs, but rose combs are sometimes seen, with red lobes and slate-grey shanks/feet. There is some variation in plumage colour, but most are gold or silver hackled blacks, as Brown-Red or Birchen Grey Game.

JURLOWSKUE GOLOSISTYE (OR YURLOWER)

Jurlow fowls are the largest of the long-crower breeds, but not the longest crowers. Most crow for about 10 seconds, which is just about long enough to keep them in this group. They have been bred for table purposes in Russia for a long time, and probably crossed with other breeds, no doubt the cause of the reduced crow.

They have been maintained at a Russian government breeding centre near Moscow at least until 2003, and probably to date. Following the sudden change from communist Soviet Union to free market Russian Federation, it is to be hoped that this and other traditional Russian poultry breeds will continue to be conserved by such official establishments. A few are now in the hands of fanciers in Germany and elsewhere in western Europe, which should prove a useful 'insurance policy' for their continued survival.

As said above, Jurlows are big birds, 3kg (6½lb) pullets up to 5kg (11lb) adult cocks. Apart from having much higher tail carriage, their body shape is somewhat oriental, like Koeyoshi. Two comb types are recognized, smallish single comb or a 'helmet-shaped' rose comb, high-domed but with a short leader. The main plumage colour varieties are: Black, Silver-Hackled Black (as Birchen Grey Game), Gold-Hackled Black (as Brown-Red Game), Silver-Grey (as Dorkings) and Salmon (as Faverolles).

JAPANESE CROWING COMPETITIONS

As the remaining breeds in this chapter are all from Japan, this is an appropriate place to give a few details of the competitions, as given by Rev. Ray Trudgian in *Fancy Fowl* magazine, August 1996. He said the contests were fairly small-scale events, mainly held in the north of Japan. Cocks were placed on stepladder-like platforms, which brings them up to eye level to their owners and the judges. There are six judges for each event, necessary because there is a lot more to it than just the length of crow, which would only have needed one person with a stopwatch.

Ray Trudgian only gave the scale of points for judging Koeyoshi; so the author does not know for sure, but guesses that a different point system is needed for each of the other breeds. With Koeyoshi, 20 points are for length of crow, 40 for volume and 40 for 'rhythm'. The birds are given five minutes to make their first crow, and are disqualified if they don't. Those, hopefully most of them, are then allowed two further 5-minute intervals to make a second and third crow. They are graded on an average score.

KOEYOSHI

These are a relatively new breed, made from the more ancient Tômaru, probably some type of Shamo-like breed, and according to the official website of the (breed homeland) Akita Prefecture Government, imported Plymouth Rocks about 1881 or 1882. Koeyoshi Fowls are well known to the local population in and around Kazuno City, not just to people interested in poultry.

A 15-second long crow is regarded as 'good', with a few exceptional specimens having a time of 20 seconds, or even a second or two more, having been recorded. Mr Ito, President of the Japanese Poultry Association is the leading expert. Some breeders, especially the very few with Koeyoshi in Europe, have to content themselves with 9-second crows.

They start their crow standing upright, then lower their head and extended neck until horizontal. Their vocal cords are well developed, as is their wide throat, with a dewlap. When they crow, the beak is nearly closed, with some of the sound coming out through the nostrils. The result is a strange, soft, bass sound.

Description

It will probably be a long time before European or American show judges will be asked to assess Koeyoshi on crowing ability. They have pea combs, ample dewlaps, thick beaks and generally show a resemblance to Yamato Gunkei. Their body plumage is quite tight, but not as sparse as Yamato, Shamo or any other of the pure game breeds. Tail carriage is horizontal, of medium length, and quite fanned. Eye and shank/feet colour are both yellow. Despite their size and apparent vigour, some European breeders have found them delicate and disease prone.

Only one plumage colour is recognized, an almost unique variation on Silver Duckwing. Males have glossy black breast, underparts, legs and tail. Neck and saddle hackles are silver with black centre striping. The wing bow, or 'shoulders' are a mixture of black, white and chestnut brown. Females have a greyish-black tail, a mostly black neck with creamy fringes and a greyish-brown breast with light feather shafts and a few dark, 'moony' spangles. Their back and wing colour is brown with black peppering and 'moony spangles' that almost qualify as lacing.

KUROKASHIWA

Translations from Japanese characters has given two more alternative spellings: Kurogashiwa and Kurokashiva. They are a black-plumaged breed with dark pigmented facial skin and a moderately long tail; which suggests a possible common ancestor with exhibition Sumatras.

The cock's crow lasts about 10 seconds, is deep, and nearer to one long, continuous note than the 'cock-a doodle-dooooooooo.....' that might be expected from a long-crower. Few details of their origin have been found by the author, beyond the fact that they come from north-western Japan and were probably developed as a distinct breed from about 1890 to 1910.

Kurokashiwa are very rare, even in Japan. In the rest of the world (in 2005), it is believed that there are only two flocks, one in Germany, the other in Italy. Most of the very few fanciers interested in long crowing breeds in the rest of Europe know each other, and they are presumably 'at the front of the queue' for any stock that can be spared from the existing two flocks. Several other rare Japanese breeds have proved susceptible to diseases and the climate in Europe, so it will probably be a decade or more before (if ever) they become a regular sight at conventional European poultry shows, or become widely available.

Long-Crowing Breeds

Description
They are not very large birds, weights ranging from 1.6kg (3½lb) pullets up to 2.8kg (6lb+) adult cocks. Combs are small to medium-sized, and single. On females, all of the comb, wattles and facial skin is dark blue-black (as Silkies). The males have dark skin on the face and the base of the comb and wattles, with the upper part of the comb and lower part of the wattles being red. Shanks and feet are black with olive soles, and more olive showing through on older birds. Kurokashiwa hens do not have very long tails, nor are the saddle hackles and main tail feathers on males anything special; but cocks do have impressive sickles and side hangers being up to 1m (3ft) long on exceptional birds. Kurokashiwa have shorter legs and a longer tail (carried at a lower angle) than the broadly similar Tômaru (*see below*) with which they might be confused.

TÔMARU

No real historical detail was found by the author, beyond saying Tômaru is one of the oldest of the long-crowers, possibly dating back over four centuries.

Their crow is usually decribed as 'baritone', but the author is tone-deaf, so will not comment further. A crow of 15 seconds is considered acceptable as far as length is concerned, with 23 seconds being the longest recorded in this breed. As said at the beginning of these Japanese long-crowers, the competitions are judged more on the required tone and notes for each breed than on length of crow.

Tômaru are similar to Mediteranean breeds in general shape, but do not have white lobes. They have medium-sized, upright, single combs and (like Kurokashiwa) have black plumage and dark facial skin, more so on females than males. Although not especially tall, their legs are longer than those of Kurokashiwa. They have moderately

Tômaru, male. Photo: John Tarren

Long-Crowing Breeds

Mr Ito demonstrating Japanese crowing with one of his Tômaru cocks. Photo: Julia Keeling

Tôtenkô, pair. The property of Mr Karia, Japan. Photo: Julia Keeling

large, fanned tails, carried well above the horizontal. Some cocks have sickle and a few side hanger feathers that are long enough to touch the ground, but the effect is not enough (that is, no long saddle feathers) to regard them as one of the long-tailed breeds. Japanese standards include a photo of a pair of Whites.

TÔTENKÔ

Tôtenko are beautiful, slim elegant birds that look similar to some German strains of Phönix, but with more open tails, slightly fanned, as Satsumadori. They have straight single combs and white ear lobes. Their true identity is only revealed when they crow!

As the fanciers who were involved in making Bergische Kräher Bantams discovered, long-crowing ability is instantly lost, probably never to be recovered, if long crowers are crossed with anything else. Tôtenko are beautiful, entertaining (but not too loud), historic and very rare; so it is vital for the few fanciers outside Japan who have them to keep them pure. They are said to be delicate in Europe, susceptible to the poultry diseases and the climate here. Only in Europe since about 2000, it will take some years of careful breeding to build up stocks here.

Tôtenkô (centre) at Shizuoka Show, Japan. Photo: Julia Keeling

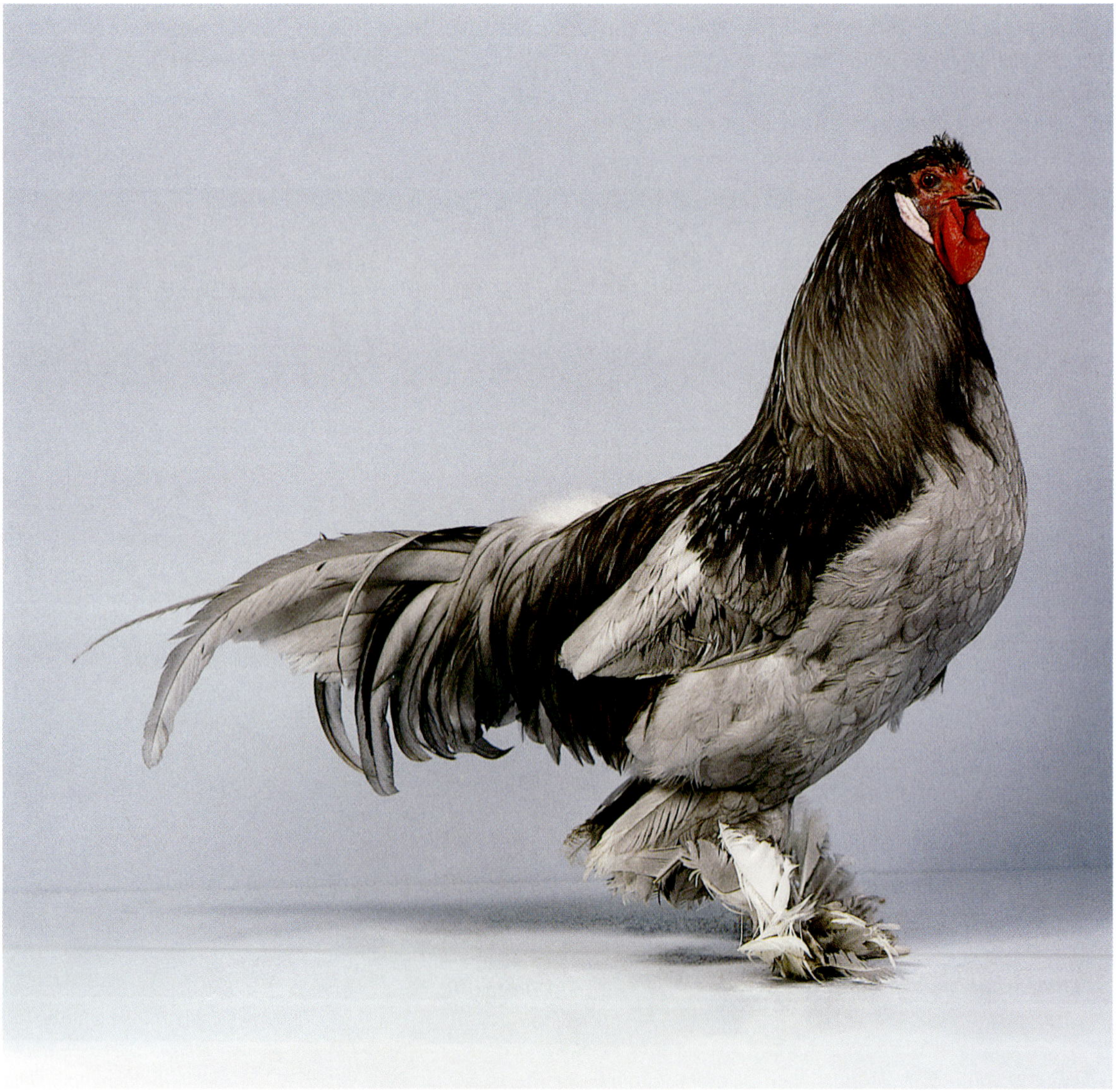

CHAPTER 6

Crested Breeds and Their Relations

The breeds without crests, but related to the crested varieties have horned combs, or in the case of Bredas, no comb at all. They also have prominent bony nostrils, a feature also seen on most crested breeds. All of the breeds described in this chapter are European in origin, in the case of the Sultan, from Istanbul, only just. However, the crested gene was probably derived from Silkies, originally from China, Japan and elsewhere in the Far East.

Some crested breeds are not covered in this book because they are popular, and so widely described elsewhere. These include: Araucana (crested British type), Barbu de Watermael/Watermaalse Baardkriel, Hollandse Kuifhoen/non-bearded Polish, Paduaner/Nederlandse Baardkuifhoen/bearded Polish, Silkie.

Crested breeds covered in other chapters of this book:

- Annaberger Haubenstrupphuhn – see Frizzled and Naked-Necked Breeds chapter
- Burmese Bantam – see True Bantams chapter
- Cream/Crested Legbar – see Autosexing Breeds chapter
- Jitokko – see Short-Legged Breeds chapter.
- Lyonnaise – see Frizzled and Naked-Necked Breeds chapter

Breeds covered in this chapter:

- Altsteirer – Austria
- Appenzeller Spitzhauben – Switzerland
- Brabançonne/Brabants Boerenhoen – Belgium
- Brabanter – Netherlands
- Breda/Kraaikoppen/Guelderiander – Netherlands
- Caumont – France
- Crève-Coeur – France
- Houdan – France
- La Flèche – France
- Nederlandse Uilebaarden/Dutch Owlbeard – Netherlands
- Pavilly – France
- Sulmtaler – Austria
- Sultan – Turkey
- Extinct Crested Breeds: Ptarmigan Fowls and Yorkshire Hornets.

ALTSTEIRER

These originated in the Steiermark region of Austria near the city of Graz. They were probably also bred in Hungary, Slovenia and Croatia. Initial development of the breed happened between 1865 and 1885, when all these now separate countries were parts of the Austro-Hungarian Empire. They were named 'Altsteirer' in 1885, a breed club was formed in Graz sometime shortly afterwards and a breed standard followed in 1894. No precise details of the matings involved have been found by the author, just a general indication that imported Asiatic fowls were crossed with local farmyard chickens. The crests would have come from these farmyard hens, and the Asiatic crossing was happening all over the poultry-keeping world in the second half of the nineteenth century. Altsteirers are not very large birds, but larger than the small, light breeds then bred in most of Europe.

Altsteirer Bantams were first suggested in 1936, but because of the Second World War they had to wait until a new miniaturization process started in 1954. Undersized large Altsteirer hens were crossed with a Welsummer Bantam cock. Wildfarbig (dark Black-Red) Deutsche Zwerghuhn (German Bantams) were added in 1958. 'Zwerg-Altsteirer' were accepted as a standard breed in 1961. The originator was H. Knöll of Klein-Umstadt.

They are only bred in significant numbers in Austria and Germany. The author did not know of any in the UK, or any outside Europe at the time of writing in 2005. When a large entry is seen at the major German shows, they are usually all from a few enthusiasts.

Description

The overall impression is of a long, horizontal-backed bird, suggesting a relationship with Dorkings. A body proportion ratio of 8:5:3 is often quoted, where '8' is body length, '5' body depth and '3' body width. Tails of Altsteirers are fairly long and widely fanned, which helps to give Altsteirers a substantial-looking build.

Combs and crests may look rather mongrelly to those not familiar with the breed, but close inspection of a large flock, or a good display at a show, reveals that they really are distinctive and uniform. Their single combs are straight on males, with the hind part curving to one side at the rear on females. The hind part of the comb on males is somewhat 'flyaway', and it is normal for them to have a lot of narrow serrations, like a (hair) comb. Judges look for broad, wedge-shaped points on most single-combed breeds, so they should remember to change their expectations when judging Altsteirers. Their crests are little more than a modest tassel below and to the rear of the combs of males, bigger

Wildbraun Alsteirer, large male. Artist: August Rixen

Wildbraun Altsteirer, large female. Artist: August Rixen

and more rounded on females. This breed has white egg shells, ear lobes, skin, shanks and feet. Weights range from 2kg (4½lb) pullets up to 3kg (6½lb) adult cocks for the large fowl, and 900g (31oz) male, 800g (28oz) female bantams.

The main plumage colour variety is 'Wildbraun', a darkish version of the Black-Red/Partridge pattern. It is similar to the well-known Welsummer colour, except that Altsteirer males have a solid black breast. Altsteirer females must have bright, light-coloured feather shafts, the same as Welsummer hens. Self White is the other currently standardized variety. In earlier times, before 1940, there were several more varieties: Barred, Black-Mottled, Blue-Red, Silver Duckwing and a lighter version of the Black-Red/Partridge pattern.

APPENZELLER SPITZHAUBEN

These are from the Appenzell Canton of north-eastern Switerland. *Spitzhauben* means pointed bonnet, and is supposed to resemble the traditional ladies' lace bonnets of the region. They are said to be the traditional breed of the region, but they are not mentioned in the old poultry books, or at least not in the English-language ones. Brabanters, the other breed with a pointed crest, are in several of the older poultry books and date back to the sixteenth century. It is not known how historic Appenzeller Spitzhaubens are. Gold Spangled Appenzeller Spitzhaubens are very similar to the now extinct Yorkshire Hornet.

However long they have existed, they were nearly extinct by 1950, and this was despite them having the advantage of originating in Switzerland, a neutral country during both world wars. Around 1953, a German breeder, Kurt Fisher of Stuttgart-Zuffenhausen, imported them and arranged for them to be accepted into the German Standards. These were in the three traditional colour varieties: Black, Gold Spangled and Silver Spangled. Since then, several more colours have been developed: Barred, Black-Mottled, Chamois (buff with spangles) and Self Blue. Silver Spangleds have always been the most popular variety.

Appenzeller Spitzhauben Bantams are a very recent creation, probably because the large breed is so small that no one thought a miniature was viable or necessary. However around 1988–90 Ehme Flemer of Aurich, in the north-western corner of Germany, made them in several colours from large Appenzellers, Ostfriessichen Zwerg-Möwe, Hamburgh Bantams and Black Rosecomb Bantams. British readers should be aware that Hamburgh Bantams in Germany and The Netherlands are much, much smaller than the Hamburgh Bantams they see at UK shows. Appenzeller Spitzhauben Bantams have not yet, as far as the author is aware, been imported into the UK.

The Hon. Mrs Pamela Jackson, one of the famous 'Mitford Sisters', made the first significant importation of them into the UK. She saw a photograph of them in a magazine when living in Switzerland in the early 1960s. The breed was very rare then, even in the Appenzell region, so it took her some time to track them down. She returned to England in 1972 with some eggs that were hatched in her sister's (Duchess of Devonshire) incubator at Chatsworth. There had been another importation, by a Swiss national then living in England that were eventually made available to the regular exhibitor/breeders.

Appenzellers were added to the list of breeds covered by the Rare Poultry Society when they started to appear at British shows, but the breeders soon (in 1978) formed a separate club, 'The Appenzeller Breed Society'. Mrs Jackson was President from the beginning until her death in 1996. Despite the enthusiasm of a few fanciers, Appenzellers never really attracted much interest among regular exhibitors, although a lot of non-showing pure-breed poultry keepers had them. The Appenzeller Society closed down, and both Appenzeller Spitzhaubens and Appenzeller Barthuhner came back into the Rare Poultry Society in 2001.

Crested Breeds and Their Relations

Gold Spangled Appenzeller Spitzhauben, large male. Photo: John Tarren

Gold Spangled Appenzeller Spitzhauben, large female. Photo: John Tarren

Silver Spangled Appenzeller Spitzhauben, large male. Photo: John Tarren

Silver Spangled Appenzeller Spitzhauben, large female. Photo: John Tarren

Description

The most obvious and distinctive characteristic of Appenzeller Spitzhaubens are their pointed head crests. The crests are more pointed on males than females. In front of the crest is a horned comb and cavernous bony nostrils. On all colour varieties, the beak is bluish, eyes dark brown, ear lobes bluish-white and shanks/feet slate-blue. General body type is similar to other light breeds, and they are one of the smallest, with pullets weighing 1.35kg (3lb) and adult cocks 2kg (4½lb).

Turning to plumage colour, the Barreds, Blacks, Blues and Black-Mottleds are self explanatory. The three Spangled varieties, Chamois, Gold and Silver, have smaller spangles than Spangled Hamburghs. These small spots do not usually impress the older generation of British judges who have spent a

Crested Breeds and Their Relations

Black Appenzeller Spitzhauben, large male.
Photo: John Tarren

Black Appenzeller Spitzhauben, large female.
Photo: John Tarren

Blue Appenzeller Spitzhauben, large male.
Photo: John Tarren

Blue Appenzeller Spitzhauben, large female.
Photo: John Tarren

lifetime looking for the almost overlapping spangles of UK-type Spangled Hamburghs. However many of the newer generation of pure-bred poultry breeders, whether or not they actually enter shows, prefer the more 'open' spots of Appenzellers. They do have one characteristic in common with Hamburghs, it is a lot easier to get neatly spangled tail feathers on Silvers than Golds. Most Golds have a lot of black shading down the length of the tail feathers.

Crested Breeds and Their Relations

BRABANÇONNE/BRABANTS BOERENHOEN

Because of the similarity of the Flemish name of this Belgian breed to the Dutch Brabanter, these two are often confused. Brabant is the name of the region of central Belgium and Noord Brabant is a southern region of The Netherlands. 'Boerenhoen' can be translated as farmyard hen, cottager's hen or rustic hen, that would be beneath the status of the Brabanter, one of the aristocrats of the poultry yard for four centuries. Brabançonne, the French name of the breed is used internationally to minimize potential confusion. This now rare breed, virtually unknown outside Belgium also has two traditional nicknames, 'Topman' in Flemish areas, 'Houpette' in French areas.

Similar-looking fowls are said (in some poultry books) to have been depicted in seventeenth- and eighteenth-century paintings by Cuip, d'Hondecoeter and others, but no specific paintings are named. They were not a 'proper breed' then, but it is very likely that birds with the small feathered tassels combined with single combs (as standard Brabançonne now) would be bred if a Brabanter cock had been running with a bunch of assorted farmyard hens.

Selective breeding started about 1890, with enough progress having been made for a breed club to be formed by 1903. In *Studie over Hoenderachtigen en Watervogels* by

Normal Quail Brabançonne, male and three females. Artist: Cornelis S. Th. Van Gink

Silver Quail Brabrançonne, male and three females. Artist: Cornelis S. Th. Van Gink

Professor A.E.R. Willems and E.T. Brandt (1971), the following founder members are mentioned: Jean Beco, A. Godtz, W. Collier, L. Michilels, G. Mignon, M. Kips and E. Keymenlen. The 1909 breed club show had 400 entries, indicating strong local interest at the time. Full details are not given, but it can be assumed that Brabançonne remained popular until the disruption of the 1914–18 war.

They began to be replaced by White Leghorns, Rhode Island Reds and other breeds that had been selectively bred for improved performance from 1920 until the 1960s, when all pure breeds were replaced by hybrids. Brabançonne hens typically lay 150–180 eggs in their first year, with the best producing about 200. This was good enough for cottage flocks, but not for commercial egg producers. Since the 1960s they have mainly been bred by hobbyists with an interest in local history.

Brabançonne Naine/Brabants Boerenhoenkrieu Brabançonne Bantam

Willems and Brandt (1971) briefly mentioned them, suggesting weights of 650g (23oz) for males and 550g (19oz) for females. They probably did not exist until the 1990s, when a strain was made using undersized large Brabançonnes and Barbu de Watermael.

Description

They are typical European light breeds in general shape, with the females being more plump and compact than the racier males.

Tails are moderately long and fairly well spread. Males have a straight single comb with a modest tassel of feathers behind. Females have modest combs that are usually somewhat twisted and over to one side. Their crests are larger and more rounded than those on males. Eyes are dark brown, ear lobes white and shanks/feet slate blue. Weights range from 1.5kg (3¼lb) pullets up to 2.5kg (5½lb) adult cocks.

The two main plumage colour varieties are Normal Quail and Silver Quail that most fanciers will know from Barbu d'Anvers and Barbu de Watermael Bantams. Other colours seen are Blue Quail, Columbian, Buff Columbian, Black, Buff, White, Laced Blue and Self Blue. The White variety has been known at least as far back as 1900.

BRABANTER

The Brabanter is a very distinctive breed, having a pointed crest (as on Appenzeller Spitzhaubens) and a tri-lobed beard. They have been bred in The Netherlands at least as far back as the seventeenth century. This

Gold Spangled Brabanter, large female, close-up of head. Photo: Hans Schippers

Gold Spangled Brabanter, large male. Photo: Hans Schippers

we know because the famous Dutch artist Melchior d'Hondecoeter (1636–95) included them in some of his paintings. 'Birds and a Spaniel in a Garden' (now in The Queen's collection, Buckingham Palace) includes a Gold Laced/Spangled Brabanter cock and a (nearly) White Brabanter hen. 'The White Peacock', apart from the main subject, has a Black Brabanter hen and other crested chickens that might be Brabanter crosses and/or Brabançonne ancestors. Apart from this artist and these two paintings, old Dutch art provides many other examples of the livestock breeds of the period, that is, everything from cows to canaries.

According to Wandelt and Wolters, in *Handbuch der Hühnerrassen* (1996), Brabanters were also bred in many of the German States circa 1806, and probably long before. Germany did not then exist as the nation state we know today. They also said that

Black and Barred Brabanters were entered in the first ever German poultry show held in 1854 at Gölitz. It is easy to imagine that the elegantly crested Brabanter would be considered the perfect breed by the middle classes who expected their birds to look good as well as lay well. Fashions change in poultry breeds as in everything else, and by 1900 Brabanters had declined almost to extinction. In Britain, Spanish suffered a similar decline over the same period. In Britain, The Netherhands and Germany, these old breeds had been replaced by Orpingtons, Wyandottes and others by 1900. There were enough Dutch Brabanter enthusiasts to save them by crossing the remaining Brabanters with Nederlandse Uilebaarden (Dutch Owlbeards), perhaps also with Laced Polish. At the present time, 2005, Brabanters are in much the same state as they were in 1900, a rare breed kept going by a few keen fanciers that sometimes need a cross to avoid inbreeding problems. The author does not remember ever seeing a Brabanter at a British show, but of course there may be some Brabanters running around somewhere in the UK.

Brabanter bantams were standardized in The Netherlands in 1934. It is believed they were made from undersized large Brabanters, Barbu d'Anvers and Laced Polish Bantams. Like the large version, Brabanter bantams are kept by a small group of keen fanciers in The Netherlands, and are virtually unknown elsewhere.

Description

The general body size and shape is typical of other European light breeds including Polish, Nederlandse Uilebaarden, Appenzeller Spitzhaubens and others. The pointed head crests are the same shape as on Appenzellers. Brabanters differ in having full tri-lobed beards and muffs. They are different from Nederlandse Uilebaarden in detail here; Uilebaarden fowls have a beard as well, but it is 'all in one', not tri-lobed.

Three of the plumage varieties are distinctive, a pattern halfway between spangling

Cuckoo Brabanter, large female. Artist: Hans Schippers

and lacing. Some Laced Polands have similar markings, broad markings at the ends of the feathers tapering out to nothing up the sides. Judges must remember that this is correct for Brabanters, they just do not have even lacing. Three colour versions are standardized: Gold/Black Laced, Silver/Black Laced and Buff/White Laced (or Chamois). Silver/Blue Laced were painted by Van Gink and Gold/Blue Laced are possible, but the author does not know if either currently exist. There are also Black-Headed Buff and Black-Headed White Brabanters, obviously related to the same colour combinations on Nederlandse Uilebaarden and Black-Crested White Polish. Other colours are: Black, Blue, Cuckoo/Barred and White.

All varieties of large and bantam Brabanters are very rare, and because the breed is some four centuries old, it is to be hoped that more people, especially outside The Netherlands, take them up.

BREDA/KRAAIKOPPEN/ GUELDERLANDER

This rare Dutch breed has three names, of which 'Breda' is the one by which they are

Crested Breeds and Their Relations

usually known around the world. They are called *Kraaikoppen* in the Netherlands, which translates as 'Crow Headed'. They do not have any combs at all, but are definitely related to the crested breeds (which is why they are in this chapter), and the main plumage variety is Black – hence 'Crow Headed'. There is another breed, called the 'Twentse' in the Netherlands and 'Kraienköppe' in Germany. Kraienköppe means 'Crow Headed' in German. British Poultry Standards name these breeds as 'Breda' and 'Kraienköppe' respectively, but if the relevent people had realized the potential for confusion at the time perhaps they would have adopted the Dutch 'Twentse', thus avoiding both versions of 'the K word'. 'Guelderiander' has also been used for them over the last 150 years or so, especially for the Cuckoo variety. Breda is a city in the south of The Netherlands near the border with Belgium and Guelderland is the region on the east of The Netherlands between Arnhem and Rotterdam.

Although never very numerous, Bredas have been mentioned in almost every major poultry book because of their unique lack of a comb together with distinctive feathered feet and vulture hocks. They are also a very historic breed and, like Brabanters and others, appear in several major seventeenth-century paintings by Frans Hals, Melchior d'Hondecoeter, Monckhorst and Jan Steen.

During the late nineteenth century Bredas were crossed with heavy Asiatic fowls like Langshans to increase their size. Bredas today have larger vulture hocks and more foot

Cuckoo Bredas, pair. Artist: Cornelis S. Th. Van Gink

Crested Breeds and Their Relations

Black Breda, large male. Photo: John Tarren

Black Breda, large female. Photo: John Tarren

Blue Breda, large male. Photo: John Tarren

Blue Breda, large female. Photo: John Tarren

feathering than depicted in the earliest pictures, possibly as a result of crossing with Sultans.

For a century and a half, Bredas have been a novelty breed, kept by those few who always choose the most unusual varieties available. No details of early British importations have been found, but Dr John C. Bennett fully described the American Breda imports in his book published in 1850 at Boston, Massachusetts. Dr Bennett quoted from a Breda/Guelderlander breeder, Mr H.L. Devereux, also of Boston:

The Guelderland fowls were imported from the north of Holland, some years since, by

Captain John Devereux, of Marblehead, in the ship *Dromo*, and since that time have been bred purely by him, at his place in that town. They are supposed to have originated in the north of Holland. They are clad in a beautiful blue-black plumage, but the flesh is white, tender and juicy. They have no comb, but a small, indented, hard, bony substance instead, and large red wattles. They are of good size, great layers, seldom inclining to sit; bright, active birds, and not surpassed, in point of beauty or utility, by any breed known in this country.

The 1905, American, edition of Harrison Weir's *Poultry Book* said that Bredas had died out in the USA by then. The author has not found any definite mentions of Bredas in the USA between 1905 and 2005. There have been some imports, usually of hatching eggs, from The Netherlands to the UK in recent years, but none have been seen at shows here since 2000.

Breda Bantam/Kraaikop Kriel

These were first made during the 1930s, but this strain died out. No doubt yet another casualty of the Second World War. Fresh strains have been made by several Dutch breeders in the 1950s, 1960s, 1970s and 1980s. It is not known how far back the current (2005) birds can be traced. Breeds used to make them included undersized large Bredas, Nederiandse Uilebaarden bantams, La Flèche bantams and Booted/Sabelpoot bantams. It has been very difficult to reduce their size, even with the Booteds.

Description

Although attention is usually focused on their head points and feathered feet, Bredas are also well-built, useful chickens. They stand fairly erect and have long backs with large, well-fanned tails. Their legs are longer than average, they need to show off their vulture hocks and feathered feet.

Returning to their headpoints, as everyone does with Bredas (30 points/100 in British Standards), they have bony nostrils as Polands and, as other crested breeds, no combs at all, white ear lobes and fairly large wattles.

Special Management Requirements

The two main considerations are their feathered feet/vulture hocks and the likely inbreeding problems with a very rare breed. A large flock kept in a lot of generous-sized houses will be needed to keep a strain going for an extended period with regular show success. Most hobbyists will not be able to keep many more breeds if they have Bredas. They are tame, friendly and fascinating birds, and so well worth keeping. Judges outside The Netherlands may not have ever seen one, but they will generally know what they are, and assess them fairly in mixed 'A.V. Rare Breed' classes.

CAUMONT

These originated in the Caumont-l'Eventé district of Normandy, France. They are related to Crève-Coeur. A breed club for the two breeds was formed in France in 1913. Caumont were mentioned as an established breed by Edward Brown in *Races of Domestic Poultry*, 1905. They were described as differing from Crève-Coeur in having a smaller crest and having normal wattles instead of the Crève's characteristic beard. Caumont are also similar to another French breed, the Pavilly. These are fully described later in this chapter. The author has not discovered exactly when Caumont became a distinct breed. Wingfield and Johnson were probably typical of books of the 1850s in suggesting that the crested French breeds had not become fully established in their standard forms. It is therefore likely that Caumont became a breed sometime during the 1880s.

Description

General body shape is similar to other European light breeds, perhaps somewhere between Black La Bresse and la Flèche. Their main feature is a pushed forward cup comb with a modest crest behind. Caumont have white ear lobes.

CRÈVE-CŒUR

Wingfield and Johnson's 1853 *Poultry Book* gives one of the earliest known references to Crève-Cœur, in a British poultry book at least. Here:

> Mr Vivian describes specimens of this bird now in his possession are of very compact form, with small top-knots, and with horned combs like those of the Polish, only very much larger. There would appear to be two varieties, since the hens in one case are represented as 'quite black;' the cock is also of that colour, but marked with yellow on the back, neck, and top-knot; the latter is a very heavy bird, 'weighing no less than 7¼lb (3.3kg); the hen being 4¼lb (1.93kg). The eggs are large and white, legs unfeathered.
>
> Of the other variety, I had given me by Lady Chesterfield a cock and three hens, black and white, somewhat after the fashion of an irregularly marked Silver Poland; they are smaller and more and more slightly made than the black birds.

This suggests some confusion between Crève-Cœur and Houdans. The same book also included a reference to 'Normandy Fowl'. The description of Normandy Fowl indicates that these were also Houdans (*see under* Houdans on p.138).

Creves never really caught on in the UK, one reason for this being that they soon gained a reputation for not doing well in our colder, damper climate. Their black plumage and dark shanks were not popular with British consumers, and therefore the specialist table bird rearers very reasonably stayed with Dorkings and related white-skinned and lighter-feathered breeds. After a brief flurry of interest during the 1870s, Crèves were reduced to the level of exotic novelty in the USA and Canada.

Harrison Weir (1902 and 1905) gave more details of Crèves in France. He quoted from two experts, M.Ch. Jaque and M.C. Lemoine. Eggs laid in winter and early spring were sent to the markets at Usieux, St. Pierre-sur-Dives, Pont l'Evêque and others. This is when prices were the highest for eating eggs. By late spring and early summer prices were much lower, so these eggs were used for hatching. Chicks from these eggs would still grow large enough for peak table bird sales from the autumn to Christmas. Later-hatched chicks of many breeds seldom have quite the bone structure of early-spring-hatched birds, but this was not a problem for the rearers or their customers. Crèves were admired for being very plump for their size. Fine bone was a bonus as French consumers did not want to spend their hard earned-cash on inedible bones.

The history of Crève-Cœur in the UK is closely linked to that of Houdans and La Flèche, and only a handful of enthusiasts were significant: S. ('Sammy') W. Thomas (died circa 1917), his son, M.R. Thomas, both of Fforest Fach, near Swansea; and George Henwood, a blacksmith from Lanlivery, a village in Cornwall between Bodmin and St. Austell. Mr Henwood's father had started keeping Crèves in 1868, and George still had some Crèves and Houdans when he was 70 years old in 1956.

Crèves usually had to compete in AOV classes. Because of the generous prize money at poultry shows in the early twentieth century, show organizers required at least twelve birds in each class to cover their costs. Here are some of the few Crève entries and results found:

> Liverpool Show, January 1892, 'AOV French', eight cocks and six hens. Second cock was a la Flèche, remaining prize winners were Crèves. Exhibitors included Mr Thomas.
>
> Crystal Palace, November 1901, 'AOV French', six cocks, six hens, seven cocker-

Crested Breeds and Their Relations

Crève-Cœur, pair. Artist: J.W. Ludlow, 1891. These birds were, 'Bred by, and the Property of, Mr J. Ainsworth, Highbank, Darwen. Winners of Cups and First Prizes at Darwen, Grantham, Liverpool, The Royal, etc.'

als, seven pullets. Two of the prizewinners were La Flèche, the remainder Crèves. Mr Thomas won cocks and cockerels, 2nd and 3rd hens, 3rd pullets.

M.R. Thomas, writing in the 1921 *Feathered World Yearbook*, said that very few Crèves or La Fléche had been seen at the shows since 1918. The largest entry found was of ten Crèves at the 1920 Crystal Palace Show. There cannot have been many more classes for Crèves provided in UK show schedules until the modern era where there is no prize money to be financed. There are very few, five perhaps, Crève breeders in the UK at present (2005), and the birds seen at the shows are small. The breed clubs are described under Houdans.

Crève-Coeur Bantam

These have been bred in the 1990s by German breeder Kurt Gemsjäger of Darmstadt-Wixhausen. They have not become very popular because it is difficult to maintain an adequate distinction between Crève Cœur Bantams and Black Poland/Paduaner Bantams. If a British breeder was going to make another strain, they would have to cross Black Poland Bantams with a heavy breed such as Australorp Bantams.

Description

There are two main considerations with these, their headpoints and general size.

Crèves are crested and bearded with a moderate-sized horned comb in front of the crest. The crest should be compact (smaller than the Polands' crests) and not obstruct the bird's vision, and the comb is much larger than the horned combs of Polands.

German Standards only require weights of 2–3kg (4½–6½lb), females and 2.5–3.5kg (5½–7¾lb), males, which is the size of the few birds seen today. British Standards still require weights appropriate for a table bird suitable for the discerning French consumer, that is 4.1kg (9lb) males and 3.2kg (7lb)

Crève-Cœur, large male. Photo: John Tarren

Crève-Cœur, large female. Photo: John Tarren

females. American Standard weights are between British and German weights. Breeders everywhere must concentrate on getting them as large as possible, or they will be virtually the same as Black Polish/Paduaners.

Plumage should be glossy greenish-black throughout. It has been said in numerous books that they were usually correct in their first adult year, but had a lot of white in crest, wings and tail in later years. Blue and White varieties have been produced since 1990.

HOUDAN

Houdans have consistently been more popular than their two close relations, Crève Cœur and La Flèche, but they have remained very much a 'Rare' or 'Minority' breed. They are named after the town of Houdan in the Department of Seine-et-Oise. As said above in the Crève Cœur section, these two breeds were only just becoming distinct and separate by about 1850. By the 1870s, Houdans in Britain were being bred very much as an exhibition breed, as was usually the case with the very show(prize money!)-centred scene here.

In their homeland they were, and among hobbyists today still are, a practical dual-purpose breed. Houdans are smaller than the other two, which means that they are nearer to their standard weights now. There was a reduction in the British standard weights for Houdans in 1923. Previously they were: cock 8–9lb (3½–4kg), cockeral and hen 6–7lb (2.7–3kg), pullet 5–6lb (2.2–2.7kg). This became 7lb (3kg) for males and 6lb (2.7kg) for females, with the comment 'not a fault if heavier'. At least since 1960 British Poultry Standards have required: male 3.2–3.6kg (7–8lb), female 2.7–3.2kg (6–7lb).

Illustrations of Houdans up to about 1880 depict birds with larger combs and plumage of an irregular mixture of black and white, often with nearly all white crests. Many Houdans in France are still like this today (2005).

The only English-language book on this breed was *The Houdan Fowl* by Charles Lee of 21 Westbourne Grove, Bayswater, London in 1874. Not many poultry breeders there now one suspects. Even the enthusiastic Mr Lee could not add much to the history of the breed, and he probably had direct contact with Wingfield and Johnson, Lewis Wright and leading French experts. Mr Lee recieved his first Houdans from 'a friend on business in Paris'. This was a trio, the cock weighing a little over 8lb (3½kg), the hens about 7lb (3kg). As was normal then, they were more white than black in plumage colour. This was in the spring of 1865, with a sitting of eggs following in February 1866. His book includes a plan of his poultry houses and runs, a typical back garden set-up occupying an area about 30 × 30ft (9 × 9m.). Mr Lee also described in great detail the force-feeding techniques used in England and France, a necessary procedure in those days to produce top-quality table birds. The last chapter of this little book comprises the first significant mention of White Houdans. He suggested that they just appeared in hatches of normal Houdans, probably lightish examples, but crossing with White Dorkings and/or White Polands. They were still a new variety in 1874.

Houdans in the USA

The first recorded importation of Houdans into the USA was in the spring of 1865 when a Mr Delrose brought some over. Many readers will remember that this was at the end of the American Civil War. Most of the progeny from the original imports were bought by I.K. Felch, one of the most famous names in nineteenth-century American poultry-keeping circles. He bred more, and shipped them into several states. The (1905) American edition of Harrison Weir's *Poultry Book* included engravings of the heads of the original type of Houdans in 1865–7. They had small and irregularly-shaped crests. Houdans were first exhibited in America in 1867 at Worcester, Massachusetts. I.K. Felch was judging and the exhibitor was John B. Gough, well known then as a temperance lecturer. From these first birds, also with a light and irregular mix of black and white plumage, American breeders started to produce their version of exhibition Houdans, slightly different from the British version detailed below.

Daniel Pinkney was the main architect of American-type exhibition Houdans. He started in 1871, and aimed from the beginning to reduce the size of the combs, increase the size and tidy up the shape of their crests. 'Several years after' [1871] he bred a cockerel with a small V-shaped or horn comb. He liked it, and mated it with several hens that still had traditional 'leaf' or 'butterfly' Houdan combs, but were smaller than normal. After a few more generations he had a strain of 'V-shaped' or horn-combed Houdans. Another leading Houdan breeder, Captain James E. White, saw them and agreed to switch to V combs. It was through Captain White's efforts that the American Poultry Association eventually switched to V combs in the Standards. No details were given, but Harrison Weir's book said that at least some American breeders argued for the butterfly/leaf/antler comb because it was the original type. At least as far back as 1890, American Standards had only required the lower weights as above, 5lb (2.2kg) pullet up to 7lb (3.2kg) adult cock.

The American Houdan Club was formed in 1898, and through the efforts of its members a display of 135 specimens were gathered at Boston Show in 1900. Most of the Houdan breeders in the USA have been in the northeastern states, only extending as far west as Wisconsin.

A few White Houdans were reported in the USA, as in the UK, back in the 1870s, but there was a specific strain developed slightly later. Mrs Leonora Hering, a famous American poultry historian, told the story in the (British) Houdan and Crève-Cœur Club Newsletter of October 1954. Mr Freeman Donald Baerman, an attorney from New Jersey, kept fancy poultry as a hobby. He started this project in 1889 by crossing a White Poland (Paduaner) cock with White Dorking and Mottled Houdan hens. It took him until 1915 to produce birds good enough to show.

Houdans in the UK

Despite the best efforts of some poultry book and magazine authors, Houdans never caught on in the UK as a commercial breed. Some may have tried them, but the history of Houdans here is essentially the story of a handful of exhibition breeders, probably the most significant being S. ('Sammy') W. Thomas and his son M.R. Thomas from Fforest Fach near Swansea, South Wales. The elder Mr Thomas was showing at least as far back as 1892 when he was at The Bath and West Show, held at Swansea that year (the current permanent site at Shepton Mallet did not exist then) and at Liverpool, one of the major shows until about 1905. He died sometime around 1917–20, with his son continuing the family tradition.

Houdans have been a very rare 'Rare Breed' in recent decades, and although they were never a major breed here, some impressive displays were once gathered:

1901 Crystal Palace Show, 9 cocks, 8 hens, 11 cockerels, 12 pullets, (total 40)
1910 Crystal Palace Show, 12 cocks, 13 hens, 8 cockerels, 8 pullets (total 41)
1911 Crystal Palace Show, 10 cocks, 15 hens, 15 cockerels, 13 pullets, (total 53)
1912 Crystal Palace Show, 8 cocks, 17 hens, 11 cockerels, 15 pullets (total 51) + 14 selling
1913 Crystal Palace Show, 10 cocks, 9 hens, 11 cockerels, 11 pullets (total 41) + 12 selling.

The original Houdan Club folded up during the 1914–18 war. A new 'Houdan, Crève Cœur and La Flèche Club' was formed in 1924. Note the difference in show entries with and without an active breed club:

1920 Crystal Palace Show, 10 Houdans in one class
1921 Crystal Palace Show, 20 Houdans in one class
1922 Olympia Show, 35 Houdans in one class
1925 Crystal Palace Show, ? cocks, ? hens, 15 cockerels, 17 pullets (total 54)
1926–39 Usually about 20 entered at the major British shows.

Crested Breeds and Their Relations

Houdans, pair. Artist: J.W. Ludlow, 1891. These birds were, 'Bred by, and the Property of, Mr S.W. Thomas, Cockett, Swansea. Cock, Winner of Cups at Crystal Palace, Birmingham, Worcester, etc.; Hen, Winner of Two First Prizes at Birmingham, Special Newport, etc.'

'French Salle D'Engraissement (Fattening Room)'. Artist: P. Colas, circa 1890. Note the force-feeding machine, used by specialist poultry table bird producers in south-east England as well as France until 1914. It was only used for the last few weeks before slaughter, and the food used was a sloppy, porridge-like mixture. It was a labour-intensive process, which explains why table chickens were a luxury product then.

There were no shows during the Second World War, or for a few years afterwards owing to fowl pest restrictions. The post-war 'Houdan, Houdan Bantam and Creve-Coeur Breeders Association' was formed circa 1953–54 and lasted, just, until the Rare Poultry Society was formed in 1969.

Houdan Bantam

W.F. Entwisle mentioned an attempt to make these about 1890, but they died out. Pringle Proud was the next to mention them, saying that he saw several good specimens around the shows in 1911. This strain almost certainly died out during the First World War. No further definite mentions of Houdan Bantams have been found until 1954. Four breeders of the bantams are mentioned in the British H., H.B., and C-C BA newsletters: T.F. Bower (Matlock, Derbyshire), Mrs Peters (Ringmer, Sussex), Eric Parker (then in Wiltshire, has moved several times since. Eric later became famous for his excellent Polands until ill-health forced him to stop) and Mrs M.S. Wiffin (Buxted, Sussex). It is not known if one or more of these people made this strain, or did they import their stock?

A German breeder, H. Schierloh, is known to have made a strain during the 1950s. He used large Houdans, Sussex bantams, Black Poland/Paduaner bantams and Black-Mottled Italiener/Ancona bantams.

No records have been found of Houdan Bantams until the 1980s. The author has not been able to find definite information on the origin of the current strains, but they are very good and deserve to be much more popular than they are.

Houdans, pair. Artist: J.W. Ludlow. This picture was originally a free gift with Poultry *magazine, circa 1912–14.*

Description

General body type is long, deep and broad, a legacy from Dorkings when they were a practical meat breed in the nineteenth century. Houdans also have five toes like Dorkings, in the case of Houdans the shanks and feet are white with black spots.

The most obvious features of Houdans are their crests, leaf/butterfly combs and beards. Lobes and wattles should be as small and insignificant as possible so as not to distract from the beard.

Young Houdans are usually almost completely black with just a few white spots. They get more white with age, especially on wings and tail. Judges should not expect Houdans to have such perfect neat white spotting as Anconas. Nevertheless, almost all white crests, as seen on White-Crested Black Polands should be avoided.

Large Houdans should weigh from 2.7kg (6lb) for pullets, up to 3.6kg (8lb) for adult cocks. British Poultry Standards (over optimistically?) require Houdan Bantams to range from 620g (22oz) to 790g (28oz). German Standards are more realistic: male, 900g (31oz), female 800g (28oz).

LA FLÈCHE

La Flèche fowls are named after a town in the Loire area of France, south of Paris. They do not have crests, although a rudimentary crest has been seen on some specimens, and are in this chapter because of their close relationship with Crève-Cœur and Houdans. They

Crested Breeds and Their Relations

Houdan, large male. Photo: John Tarren

Houdan, bantam female. Photo: John Tarren

may also have an ancestral relationship with Brabanters, Bredas and Nederlandse Uilebaarden/Dutch Owlbeards.

Harrison Weir (1902 and 1905) said that this is an ancient breed, dating back to the fifteenth century. He also quoted from a sixteenth century reference to them, Maister Prudens' book *A Discourse of Housebandrie* (1580): 'that in the choice of the cock you shall consider the plumage of feathers; the black, red, or tawnie are the best; also that they have their combs crests up-right, and double or divided.' As he recommended that the best birds came from Angeow, Touraine and Lodumoys, this supports the idea that these were the ancestors of La Flèche. No further details of their history before 1850 have been found by the author.

They were reared commercially in their home region until the late nineteenth century, but Edward Brown reported that when he saw them (researching *Races of Domestic Poultry*, 1905) in the town of La Flèche and the surrounding La Sarthe region they were only bred by amateur poultry keepers. Although they still had a reputation for being the finest quality table chickens, they were too slow growing for normal market purposes. Mr Brown met a few people who were still rearing them in the traditional way for the dead stock classes at agricultural shows.

'The traditional way' of rearing them for the table meant that they were not slaughtered until eight, sometimes ten, months old. Compare this with today's broilers, killed at a similar number of weeks (not months) for similar oven-ready weights. Around 1900, when Mr Brown was doing his research, killing ages between four and six months would have been normal for Indian Game × Faverolles. Traditionally reared La Flèche were out free-ranging, finding worms and insects as well as vegetation, until the last few weeks of intensive feeding of damp mashes; milk and breadcrumbs to start, then buckwheat, ground maize and oats with water. Fattened (and caponized?) cockerels sometimes reached 5.5kg (12lb) and fattened pullets 4kg (9lb). Potential breeding stock, not fattened, were usually about 1kg (2lb) less.

La Flèche have only been kept by a few enthusiasts outside France, with virtually no details having been found of them in the USA. There is at least some history worth recording of them in the UK and Germany.

La Flèche in the UK

The Hon. C.W. Fitzwilliam was the leading British breeder in the 1860s and 1970s. Others tried them, but most gave up when the

Crested Breeds and Their Relations

chicks failed to thrive, or even died, in our colder, damper climate. George Henwood, also mentioned in the Crève-Cœur section, bred both breeds for most of the first half of the twentieth century. He recalled, in the 1930 *Feathered World Yearbook*, 'The late Mr C. Isherwood and I had a chat on La Flèche at Bodmin Show in 1928, and he told me his father kept La Flèche when he was quite a lad.' This would take it back to about the 1850s or 1960s.

Mr Henwood had earlier (1927 *FWYB*) complained about the 'Houdan, Crève-Cœur and La Flèche Club': 'La Flèche is only a word put on the club heading. Not a class has been put on since it was formed' (circa 1924). The 1931 *FWYB* reported that Mr Henwood was the only remaining La Flèche exhibitor in the UK, and had been showing them for more than thirty years. He at least had the satisfaction of winning some 'AV French' classes, such as at the 1926 Bath and West Show. These UK stocks had probably died by 1945, with present La Flèche in Britain being imports from France or Germany.

La Flèche in Germany

More La Flèche have been bred in Germany than anywhere else outside France. Dr Lax of Hildesheim founded a breed club for them in 1859. German fanciers accepted lower weights than those of France and Britain. Another difference was that in these two countries they only ever had the Black variety, whereas the German breeders developed Blues, Cuckoos and Whites as well. During the 1920s and early 1930s, some of the German La Flèche Club Shows had over a hundred entries.

White La Flèche, pair. Artist: Kurt Zander

The breed had to be revived after the war, the main breeder being K. Fischer of Stuttgart-Zuffenhausen, who started this process in 1949. By 1953 he had showable birds made from Augsburgers and other breeds. It is not known if he managed to find any survivors from the pre-war strains.

La Flèche Bantam

These were first made in Germany in the 1960s from under-sized large fowl, Deutsche Zwerghuhner (German Bantams) and Rheinländer Bantams. There were enough fanciers to exhibit thirty-five at the 1977 Karlsruhe-Knielingen Show and a splendid display of fifty-one at the 1996 Sprendlingen Show. The author assumes that some also exist in France by now, but is not aware of any elsewhere.

Description

The general body shape of La Flèche is more Mediterranean than the stockier Crève-Cœur and Houdans, even more apparent with the slimmer German strains. German Standards only require a weight range from 2kg. (4½lb) pullets up to 3kg (6½lb) adult cocks, whereas British and French judges would hope to find them at least 1kg (2lb) heavier. They were originally intended to be a table breed after all. Bantam males should weigh 900g (31oz), females 800g (28oz).

Horned combs are the defining characteristic of La Flèche, and at the base of the comb there are cavernous, bony nostrils, a clear indication of their relationship with crested breeds. They have white ear lobes of a fair size, red or brown eyes and, unlike Crève-Coeur and Houdans are not bearded, instead having long wattles. Young Black La Flèche have black, or nearly black shanks and feet, fading to slate-blue with age; the latter being the shank/foot colour of Blues of all ages. Cuckoos and Whites have white shanks/feet.

Breeders should concentrate on obtaining maximum size if birds will ever be seen again to compare with those described and illustrated in the classic old poultry books.

Black La Flèche, large male. Photo: John Tarren

Black La Flèche, large female. Photo: John Tarren

NEDERLANDSE UILEBAARDEN/ DUTCH OWLBEARD

Dutch Owlbeards were depicted, as were Brabanters and other historic breeds, on seventeenth-century 'Dutch Master' paintings. Unfortunately, this ancient breed declined during the second half of the nineteenth century, pushed aside by Asiatic breeds like

Crested Breeds and Their Relations

Moorhead Dutch Owlbeard / Nederlandse Uilebaard, male and four females. Artist: Cornelis S.Th. Van Gink

Langshans, and part Asiatic, imported new breeds like Orpingtons, and eventually the part Asiatic local creation, the Barnevelder. Neglect of the grand old Owlbeard was not helped by it not being officially standardized until 1882. By 1900 they were nearly extinct, only to be saved by the formation of the 'Nederlandse Hoenderclub', devoted to the preservation of the ancient Dutch breeds. Its members found all the surviving Owlbeards and potentially useful Owlbeard crosses, mating some of them with La Flèche and Thüringian Beardeds to solve the inbreeding problems of the pure-bred birds. Dutch poultry expert, Cornelis S. Th. van Gink reported in the 1914 *Feathered World Yearbook* that:

The Owl-bearded Dutch fowls have improved a good deal in the last three years, their colour having been much perfected. They are hardy birds, and will stand a good deal of cold. In type they are somewhat like a short-backed Hamburgh, but with a higher tail. Their beard is full and round, and should not hang down so much as in Faverolles. Their comb is a two-horned La Flèche comb, which gives their heads a fine appearance. The principal colour still is Black, in which variety some capital birds are always shown, but Whites are fairly popular too. Besides these self-colours, there are Silver-spangled, Gold-spangled and Gold-pencilled varieties. The males of the Spangled varieties have black breasts and a black beard, and the females also have a black beard. A very fine Buff male at the late [1912? 1913?] Utrecht Show, of good type and colour.

Crested Breeds and Their Relations

Black Dutch Owlbeard/Nederlandes Uilebaard, large male. Photo: Hans Schippers

White Dutch Owlbeard/Nederlandse Uilebaard, large male. Photo: Hans Schippers

Since then some additional varieties have been developed, with the emphasis on those that existed in the seventeenth, eighteenth and early nineteenth centuries. There have also been changes in the Spangled varieties; they now have spangled breast feathering, not black as in 1913. Two new Spangled varieties have been created, buff with white spangles and silver with lavender spangles. Photographs of current (1995–2005) strains show variation in the size and shape of the spangles, from small round spangles (as Appenzeller Spitzhauben), to half moon spangles, approaching lacing (as Brabanters). Gold and Silver Pencilled Owlbeards are standardized, with markings similar to that of pencilled Friesians. In self colours, there are still Blacks and Whites, now plus Blues (laced, but not as sharply as the best Andalusians) and Cuckoos.

There is another pattern, the 'Moorkops', available in a range of colour combinations. This name translates as 'Black-Heads', and they are certainly related to Black-Crested White Polish, possibly also to Lakenvelders. These varieties are: Black-Headed White, Black-Headed Golden-Yellow and Black-Headed Golden-Brown. Blue-Headed versions of all these have been attempted. The ideal is a solid black (or blue) head, with the colour extending partway down the neck, then stopping, to give remaining body plumage as clear as possible. Young birds have a lot of black in body feathering that usually disappears when they get their adult plumage.

Owlbeards remain rare, although there are some enthusiasts in Germany as well as The Netherlands. A few have been seen in the UK since 2000, but it probably needs a few UK resident fanciers of Dutch or German origin with real loyalty to Owlbeards to secure their position here.

Nederlandse Uilebaardkriel/ Dutch Owlbeard Bantam

These were only recognized in The Netherlands in 1935, and had to be remade in the 1950s. They are larger than one would like, considering the modest size of the large fowl. Comb shape and plumage pattern also require further improvement.

Description

General body size and shape is similar to that of Polish or Hamburghs, with large weights from 1.6kg (3½lb) pullets up to 2.5kg (5½lb) adult cocks; the equivalent bantam range

Crested Breeds and Their Relations

Blue Dutch Owlbeard / Nederlandse Uilebaard, large female. Photo: John Tarren

Moorhead Dutch Owlbeard / Nederlandse Uilebaard, large female. Photo: Hans Schippers

being 600g (21oz) to 800g (28oz). The horns of the combs should match and be large enough to 'look the part'. Some Owlbeards have rudimentary crests, which is OK, as long as they remain rudimentary.

New fanciers might decide to keep two colour varieties, one of the straightforward self colours for comparatively easy showbird production, and one of the spangled or moorkop varieties, for a real conservation challenge. Breeders should work as co-operatively as possible, this depending on how far apart they live.

PAVILLY

Edward Brown mentioned (*Races of Domestic Poultry* 1905) Pavilly as an alternative name for Caumont Fowl that are also described in this chapter. Pavilly is a small town northwest of the city of Rouen in the Département of Yvelines, of the Haute-Normandie Region. The fowls remain very similar to each other, with the most obvious difference being cup combs on Caumont and single combs on Pavilly, in both cases, combined with a crest. It would be very difficult, probably impossible, to find any older history about either breed. Before 1850, both of them, plus Crève-Cœur and Houdans could all be lumped together as 'Normandy Fowls'.

There is only one standard colour variety, Black that has slate-blue shanks and feet. Now a medium-sized breed (2kg (4½lb) pullets up to 3.5kg (7¾lb) adult cocks), they were much heavier when bred by specialist table bird fatteners. Pavilly fowls are likely to remain virtually unknown outside France.

SULMTALER

These originated near the river Sulm that flows to the south of the city of Graz in the Steiermark region of south-eastern Austria. Sulmtaler fowls were developed during the late nineteenth century from Altsteirer, the older breed of the area, and several larger breeds (Brahmas, Cochins, Dorkings, Houdans and Langshans) to increase size for meat production. Amin Arbeiter of Feldhof is credited with being the main breeder responsible for making a uniform breed out of this mixture by about 1905. They became a popular table breed in Austria, Germany and, to a lesser extent, The Netherlands, from 1920 to 1940. Post-war, they were revived in Austria and both East and West Germany, obviously declining somewhat when hybrid broilers came in for commercial use.

Crested Breeds and Their Relations

Sulmtaler, large male. Photo: John Tarren

Sulmtaler, bantam male. Photo: John Tarren

Zwerg-Sulmtaler/Sulmtaler Bantam

H.-J. Webers of Isemhagen, Germany made Sulmtaler Bantams during the 1950s. He entered them in the Hanover Youngstock Show, circa 1957–60. Since then they have become more popular than the large fowl. A few were imported to the UK about 1990, which attracted the interest of a few, mostly lady, exhibitors.

Description

The ratio of 3:2, depth (back to belly) to width (across the shoulders) is the recognized ideal proportion of a Sulmtaler's body, with a probable body length (excluding tail) ratio of 4 or 5. They are therefore well built birds with a straight back and a full-rounded breast.

Sulmtaler males have a straight single comb, inclined to be flyaway and with a lot of narrow serrations, plus a small crest or tassel of feathers at the rear. Females have larger crests, and combs that are 'S'-shaped when viewed from above. This curve is called 'Wickel' in German. Both sexes have orange-red eyes, small/medium white ear lobes and normal wattles. Beak, skin and leg colour is pinkish-white.

Sulmtaler, bantam female. Photo: John Tarren

For most of their history the only colour variety was Black-Red/Wheaten, with a new Blue-red/Blue-wheaten variety being developed since 2000. Black-Red males are a bright shade, with neck hackles more gold than the richer red shades of the saddle. A

Crested Breeds and Their Relations

little black centre striping is allowed in hackles, but only a little. Similarly a few, but only a few, brown feathers are allowed in the ideally black male breast plumage. Sulmtaler females have pale, creamy crest, breast, belly and leg feathering, a rich orangy neck and the back and wings are brown, with a little black shading. Most British fanciers and judges seem to like the females a lot more than the males.

SULTAN

Most old breeds of poultry, including many of the crested breeds in this chapter, were once practical layers or table birds. This was never the case with Sultans, they were always a purely ornamental breed. They were introduced to the Western world by Miss Elizabeth Watts of Hampstead, London, and this is the story of their importation, in her own words.

From *The Poultry Yard*, by Miss Watts, circa 1870: 'The Serai-Taook, or Fowls of the Sultan':

They were sent to us by a friend living in Constantinople [now Istanbul], in January 1854. A year before, we had sent him some Cochin-China fowls, with which he was very much pleased; and when his son soon after came to England, he said he could send from Turkey some fowls with which we should be pleased. Scraps of information about muffs, and divers beauties and decorations, arrived before the fowls, and led to expectations of something much prettier than the pretty ptarmigan, in which we had always noticed a certain uncertainty in tuft and comb.'

[Refers to Ptarmigan Fowls, see the end of this chapter.]

In January they arrived in a steamer chiefly manned by Turks. The voyage had been long and rough; and poor fowls so rolled over and glued into one mass with filth were never seen. Months afterwards, with the aid of one of the first fanciers in the country, we spent an hour in trying to ascertain whether the feathers of the cock were white or

Sultan, large male. Photo: John Tarren

Sultan, large female. Photo: John Tarren

striped, and almost concluded that the last was the true state of the case, although they had been described by our friend as 'Bellissimi galli bianchi.

Miss Watts had to wait until the birds moulted before she could see how beautiful, and pure white they were. She wrote to her friend, to ask if he could obtain more, but he could not. Only one hen, imported by a Mr F. Zorhorst many years later, was recorded as coming from Constantinople. Successive generations of Sultan enthusiasts have managed to keep Sultans going through the next 160 years to the present. When inbreeding has brought strains to a standstill, a Houdan or White Poland/Paduaner cross has revived things, athough a few generations of careful selective breeding have been needed to restore foot feathering and other breed characteristics.

One of the leading Sultan breeders of the early twentieth century was Mrs Campbell, mainly remembered in poultry history as the creator of the Khaki Campbell Duck. In the 1927 *Feathered World Yearbook*, she said she was then 'the longest consecutive breeder of them in England, as I have reared some every season for over thirty years', that is, since about 1895. In the 1924 *FWYB* she commented that many of her visitors were very excited to at last see live specimens of a breed they had previously only seen pictures of. Some things never change in the poultry fancy world, enthusiasts today are still excited to see breeds, recently imported perhaps, of which they also have only ever seen pictures.

From their arrival in the 1850s, until the foundation of the Rare Poultry Society in 1969, Sultans have almost always had to compete in AOV classes; but such is their beauty and individuality, that the few entered have usually come away with one of the prizes. This is still the case where they are in mixed Rare Breed classes, but with present minimal prize money, there are more shows with separate Sultan classes than ever before.

Sultans have generally only ever existed in one colour variety, White. However in 1929, Miss Lambert bred a Cuckoo, illustrated in the 1930 *FWYB*. In the article with this photo, it was said that someone (not named) was working on Blacks. Another prominent Sultan breeder, also famous in her time for Silkies, Mrs Fentiman of Swindon, Wiltshire, confirmed in the 1935 *FWYB*, that this was the one and only Cuckoo Sultan ever bred, and that the Blacks never appeared in public. Mrs Fentiman managed to keep her strain going through the Second World War, at least up to 1956. There may have been others in this period, but not many.

No further details have been found until about 1965, when there were two known breeders, Eric Parker, better known as the leading Poland exhibitor in the UK for several decades, and F.E.B. Johnson, owner of Stagsden Bird Gardens, a small private zoo in Bedfordshire; each then unknown to the other. Eric passed his Sultans on to Rare Poultry Society founder, Andrew Sheppy (about 1968?), after he decided that crests, feathered feet and white plumage, when all on one bird were too difficult, for even he (a master at show preparation techniques) found this too much. Some of Mr Johnson's birds were bought by Richard Billson, later RPS Secretary, and the two strains were crossed by both Andrew and Richard, to alleviate inbreeding problems.

Moving on to the present (2005) there are about ten recognized flocks in the UK (regular exhibitors and farm parks), with an estimated 200 breeding hens between them; plus, no doubt; some unknown flocks as well. There are also Sultans in the other major poultry showing countries, including, now, some Blacks and Blues.

Sultan Bantam

Surprisingly for such a long-established and ornamental exhibition breed, very few Sultan Bantams have appeared over the last 150 years, with most of the few being in the USA. Anyone attempting to make a strain would probably start by crossing undersized large Sultans with White Barbu d'Uccles.

Description

Being crested and bearded, one might assume that their headpoints would be identical to those of White Polish/Paduaners, but there are noticeable differences. The horned combs of Sultans are larger and more defined than the usually small and indistinct horned combs of Polish. Sultans' crests are more compact, rounded and swept back than those of Polish, especially so on males. Wattles and lobes are as small as possible, and mostly covered by the beard.

Large Sultans do not have the desired deep body, with a full, rounded breast until two years old. When most other breeds are ready for their first show, say about eight months old, Sultans are still 'gangly teenagers', with flat or 'cut-away' breasts. Carriage, described as 'sprightly', is more apparent on males who stand more upright than the more sedate-looking females who have an almost horizontal back line. Body plumage is abundant and profuse, which makes them look larger than other breeds of the same weight. Standard large fowl weights range from 2kg (4½lb) pullets to 2.7kg (nearly 6lb) adult cocks. The equivalent weights for bantams are 510g (18oz) and 790g (28oz). It would probably be difficult to get Sultan Bantams down to these weights.

Apart from the obvious heavy leg and foot feathering with large 'vulture hocks' (long, stiff feathers projecting down and back from the hock joints), Sultans also have an extra hind toe that should be clear and distinct from the normal hind toe, be longer than it, and point up, clear of the ground. Beak, shanks and feet are all white or pale blue, which along with the five toes, crests and beards, all suggest that Sultans are related to Silkies. There were trading links between China (Silkies) and Turkey (Sultans) for many centuries.

Special Management Requirements

As Eric Parker would have said, 'All of them!' See *Exhibition Poultry Keeping*, also by David Scrivener, and also published by The Crowood Press, for details of the techniques required. Successful Sultan exhibitors will probably not have many other 'high maintenance' varieties.

EXTINCT CRESTED BREEDS – PTARMIGAN FOWL AND YORKSHIRE HORNET

PTARMIGAN FOWL

These were similar to Sultans, but slimmer and tighter feathered. They arrived in England about a decade before Sultans, and probably died out during the 1860s. Ptarmigan Fowls were described and illustrated (with a wonderful chromo-lithographed plate) in *The Poultry Book* (1853) by Rev. W. Wingfield and George W. Johnson, both of whom also produced *The Cottage Gardener* magazine, the other source of information on this long-gone breed. The birds in the picture (from a water colour by Harrison Weir) were tall and slim, had long vulture hocks with (unlike Sultans) quite modest foot feathering. They were not bearded, and had modest crests, which on the cockerel were pointed like those of Appenzeller Spitzhaubens and Brabanters.

The Cottage Gardener magazine had two relevant articles, extracts of which are given here: The 4 August 1853 issue included a report of the 'Summer Metropolitan Poultry Show' held at the Baker Street Bazaar:

> The greatest novelty here were the Ptarmigan or Grouse-footed Polands, exhibited by Dr Burney of Brockhurst Lodge, near Gosport. The old birds are almost as small as Dumpies; white with slightly coloured hackle, white top-knots, and remarkably well-feathered or booted legs. The combs are cupped, and the cock's tail is well-sickled.

The 26 December 1854 issue included a letter from 'W.H.' that gave more information than Dr Burney was able to provide:

> Some fourteen years, or nearly, ago, Mr. J.E. Elworthy, of Plymouth, had four fowls sent him by a friend, an officer in the Royal Navy, which had been bought at Constantinople. These fowls were much admired by those who saw them, and Mr. Elworthy gave chicken and eggs to many of his friends.

This letter went on to give details of how, via several others, some of this strain became the property of Dr Burney. He continued with details of further importations:

> ... during the past year, several lots of the same description have been recieved from Constantinople and the Bosphorus. Captain Russell, the commander of the celebrated steam ship, *Himalaya*, has brought over some procured at Constantinople, clearly of the same variety as Dr Burney's; and a gentleman of Devonport, who sent out a friend of his, in one of the ships on the Bosphorus, but not a fowl-fancier, to send him home some of the fowls of the country, and has received the same variety.

YORKSHIRE HORNET

The only known photos of Yorkshire Hornets, one each of a cock and a hen, appeared in the 14 November 1930 issue of *Feathered World* magazine, and were taken by Arthur Rice. The birds belonged to Mr F.C. Branwhite of Otten Belchamp Hall, Clare, Suffolk, who, in a letter sent in with the photos, said his family had kept Yorkshire Hornets for over eighty years, which would take this otherwise little known breed back to about 1850. These birds were generally similar to Gold Spangled Appenzellers in terms of plumage pattern and pointed head crest, but said to have been considerably larger. Mr Branwhite said (in 1932 *FWYB*) that cockerels could be fattened up to '6 to 8 lbs'. (2.7–3.6kg).

Poultry breeds were not all as fixed and widely identified as they are now, in the 1840s and 1950s, as confirmed (in this case helpfully) in two American books: Dr John Bennett's *The Poultry Book* (1850) and D.J. Brown's *The American Poultry Yard* (1857). Both contain engravings of 'Spangled Hamburghs' – with crests.

It is not known when or how the breed first became established in Yorkshire. Were they made there? (by crossing Spangled Hamburgh-type fowls with crested fowls), or were they imported more or less ready made, as Brabanters perhaps, and then slightly modified and renamed?

The last known Yorkshire Hornet, an aged and infertile cock, died about 1970. It was old when Andrew Sheppy (RPS founder), of Congresbury, near Bristol, got it. The previous owner did not have any hens (or any more cocks) left, so Andrew tried, without success, to at least get some crosses (with OEPF for Redcaps), but it was not to be.

It would probably not be worth remaking Yorkshire Hornets. Any effort in this direction at present would be better spent on Appenzeller Spitzhaubens and Brabanters.

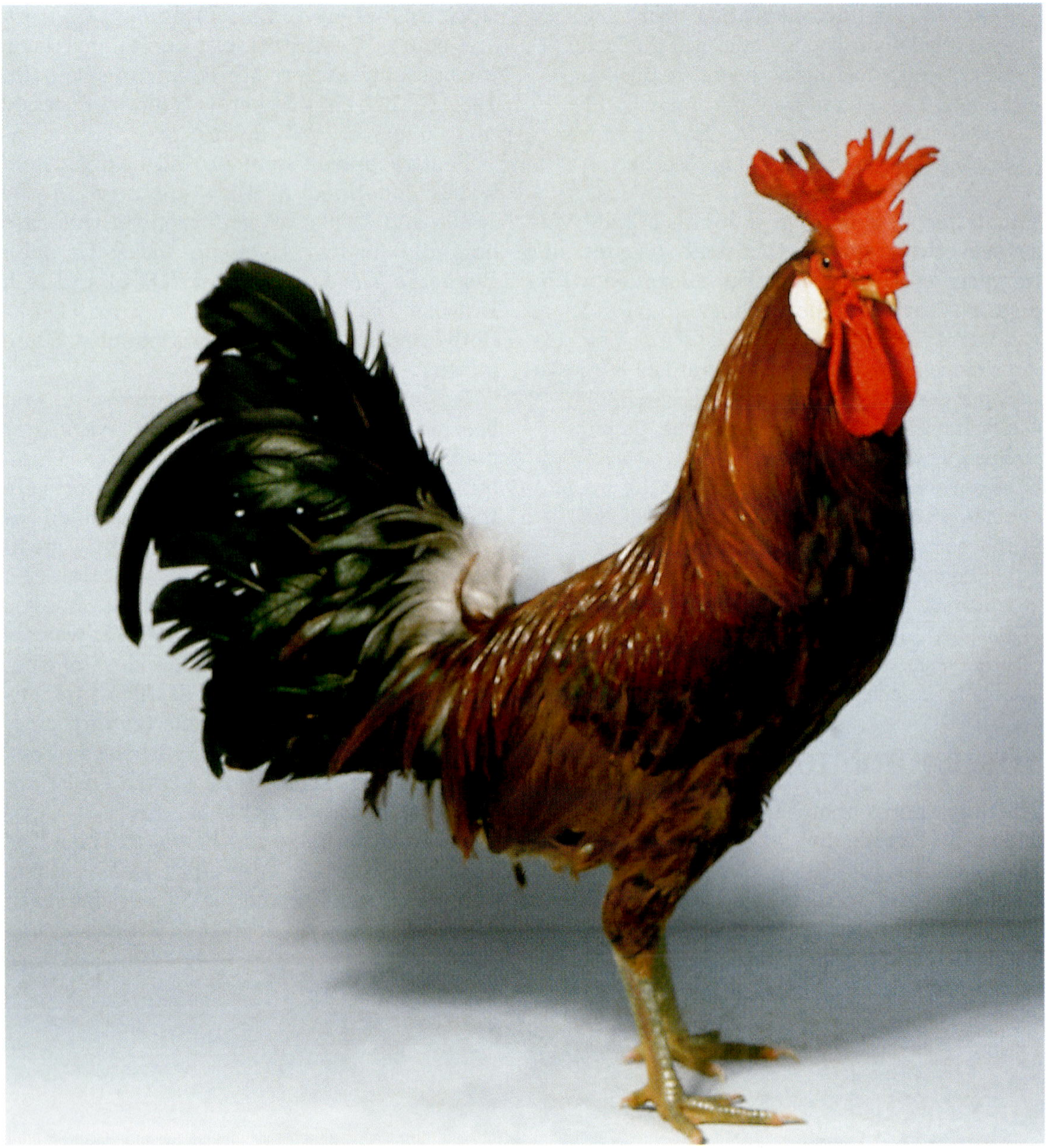

CHAPTER 7

Cup-Combed Breeds

This chapter covers: Sicilian Buttercup (Sicily, then USA), Sicilian Flowerbird (believed extinct), Coveney White/White Coronet (UK, extinct), Augsburger (Germany), *N.B.* Caumont (France) are in the Crested Breeds chapter.

Cup combs are caused by the Duplex gene acting on a single comb to divide it into a crown-like structure. The gene is very variable in its effect, with many having nearly a normal single comb with just a tiny cup at the rear. Only a small proportion of each batch have show quality combs, probably a reason for their limited popularity.

As with many other poultry breeds, there is not much documentary evidence of their existence before the beginning of showing in the nineteenth century, however the ancestors of Sicilian Buttercups and the others certainly existed in the Mediterranean area long before then. The last edition of Wright's *Book of Poultry* (revised by S.H. Lewer) mentions an article in the 1914 American Buttercup Club's Yearbook by Mrs. A.N. Dumaresq. In this, she said that Buttercup-type birds could be seen on paintings, some dating back to the sixteenth century, hanging in the major art galleries of Florence, Paris, Rome and the Vatican. Only a painting by Spanish artist Spinello in the Vatican Gallery is specified.

SICILIAN BUTTERCUP

Although the original stock came from Sicily, they were stabilized and widely recognized as a breed in America. They are said to have first arrived in the States with Sicilian immigrants in the 1830s, but no details have been found. The first documented importation was in 1860, when Captain Dawes brought some on his barque (a specific type of sailing ship) *Fruiterer* on one of his regular trips from Sicily to Boston. The chickens were on board to provide fresh eggs and meat on the long journey across the Atlantic; and had laid so well that the remaining (that is, those not eaten!) hens were taken to Captain Dawes father's farm. Mr C. Carroll Loring of Dedham, Massachusetts bought them from the farm shortly afterwards, and liked them so much that he arranged for more importations over the next few years. He bred, advertised and promoted them as best he could over the next fifty years, but without attracting widespread interest.

Finally, in 1908, Mr J.S. Dumaresq of Easton, Maryland and his friend, Mr L.B. Audinger, publisher of the *Industrious Hen*, Knoxville, Tennessee took them up, being better able to promote them than Mr Loring. The American Buttercup Club was formed in March 1912. By 1914 it had about 500 members over America and Canada, a much larger membership than many longer established breed clubs anywhere in the world. This level of interest could not last, and it did not. Most of the large scale breeders were concentrating on White Leghorns, so their egg production soon surpassed every other breed. Progress was much slower with Buttercups. Apart from anything else, many of the early birds

Cup-Combed Breeds

Golden Sicilian Buttercups, pair. Artist: Ernest Wippell. Ideal comb shape and plumage markings are easier to draw than breed in real life.

could only be identified as such by their cup combs and approximate size and shape. The plumage pattern was very variable, although most had some type of black markings over a gold ground colour. Present-day breeders of laying hybrids do not select for specific plumage colours and patterns, but back then farmers expected a flock of a 'pure breed' to look reasonably uniform. The agreed Buttercup pattern was admitted to the American Standards in 1918, and is described below. It is still the only colour variety of Sicilian Buttercup recognized in the US, despite a wider range of colours being introduced when the breed arrived in Britain. No further details of the rise and fall of show entries and general popularity in the US were available at time of writing.

Sicilian Buttercups were brought to Britain from America, not Sicily, about 1911 or 1912. A British Sicilian Buttercup Club was quickly formed, shortly followed by an excellent eighty bird entry at their first Club Show as part of the 1913 Crystal Palace Show. The judge for these Buttercups was Mr. Ernest Wippell, a well-known name among today's (2005) collectors of old poultry prints, books and other ephemera. Wippell did several paintings of Buttercups, including some given as free gift prints with *Poultry World* magazine: a cockerel with the 21/11/1924 issue, a pullet following in the 9/1/1925 issue, also a pair on an unknown date, possibly circa 1921, and they appeared on several multi-breed pictures. Such was his influence at the time, one is inclined to wonder if his pictures were at least partly responsible for the differences in British Standards in ear lobe colour and plumage pattern from the established American Standards.

The annual Sicilian Buttercup Club Shows (usually at Crystal Palace) had entries of up to 100 birds for the first few years after the First World War, but a sharp decline was evident in 1926, when just twenty-seven were on display. The original Gold Buttercup was in an even worse state of decline because many of these twenty-seven Buttercups were of new colour varieties that had been developed by British fanciers. In both the UK and the USA, in the hard times of the Depression, both commercial white egg producers and light breed show exhibitors were playing safe, which usually meant Leghorns, sometimes Anconas or Minorcas, rather than Buttercups.

From the original 'Gold Buttercups', the next variety was the Silver, the same black markings as described in detail below for Golds, but on a white ground. Brown Buttercups were developed by Mr Chris Isherwood about 1915 and were recognized by the PCGB in 1920. Browns were similar in colour and pattern to Brown Leghorns or Black-Red/Partridge OEG Mr Isherwood teamed up with Mr R. Terrot to cross Silvers with Browns to

Golden Sicilian Buttercup, large male. Photo: John Tarren

Golden Sicilian Buttercup, large female. Photo: John Tarren

produce Duckwing Buttercups (as Duckwing Leghorns or Game) around 1920. Black Buttercups were first exhibited in 1928 by a Mr Hardcastle of Langworth, Lincolnshire, who was also Secretary of the Sicilian Flowerbird Club (*see below*). Only Golds and Silvers survived through the Second World War to the modern era, with RPS Secretary Richard Billson being one of the leading breeders. The British Sicilian Buttercup Club closed down, never to be revived, during the war. Black Buttercups have now effectively been replaced by Augsburgers (*see below*).

Sicilian Buttercup Bantam

These were admitted into American Standards in 1960, but the author has not discovered how many existed then, or who made them. British Standards have given suggested weights for them long before any existed here. The author first saw some, two Gold pullets, when judging at the Suffolk and Essex Poultry Club Show in October 2004. More selective breeding seemed necessary to obtain birds that would remain small bantams as adults.

Description

Very few Sicilian Buttercups have anything approaching ideally shaped cup combs. They are expected to have at least one normal spike (as on single combs) at the front before the comb divides to give the crown or flower effect. Many birds have several spikes as a single comb, with just a modest cup at the rear. Most paintings of Sicilian Buttercups show upward pointing comb spikes forming a crown-like effect, but this was usually what the artist would like to have seen, rather than what was seen. Photos of most show winners show combs with nearly horizontal spikes, giving a flower or star-like effect. To be really perfect, the comb should be 'closed' at the rear, to give a more definite cup or crown-like effect; but this rarely appeared.

Buttercups are required to have white ear lobes in American Standards and red ear lobes (up to a third white allowed) in British Standards. The reason for this difference is not clear, but the author's best guess is that the American authorities thought white lobes correct because all the other Mediterranean breeds (Anconas, Leghorns, and so on) had white lobes. British breeders may have had stock with indistinct (half white/half red) coloured lobes, and wished to avoid the difficulties of maintaining white lobes. They may have forseen the difficulties they would have just getting the combs and plumage pattern right, and did not want to give themselves another problem.

General size and body shape is typical of Mediterranean breeds, with large fanned tails required. British Standards require large males to weigh 3kg (6½lb), females 2.5kg (5½lb), bantam males 735g (26oz), females 620g (22oz). On all colour varieties, the beak is dark horn with yellow shading, eyes are red-bay, shanks and feet willow-green.

Gold male plumage pattern and colour: the general effect is of a rich orange-red bird with a glossy black tail. There are some black spangles at the base of the neck and around the thighs. Main primary and secondary wing feathers are bay on the outer web, black on the inner. Some of the lesser sickle tail feathers are black with gold edging, similar to ideal tail feathers of Gold Pencilled Hamburgh males. The most common fault has been washed-out ground colour, sometimes even with white patches or white edging on breast feathering.

Gold female plumage pattern and colour: this is very different from their mates, with a golden-buff ground colour, on which all but the breast feathers have a delicate 'ears of wheat' pattern of black spots. Although both American and British Standards require the neck and upper breast to be clear golden-buff with the lower breast and belly spangled, the idealized paintings and photographs of winning birds show that the reality was slightly different. British Buttercup hens seem usually to have almost clear fronts right down to their legs, wheras American hens were often shown with spangling well up the breast.

Silver Buttercups have the same black

markings on a pure white ground colour. Browns and Duckwings are no longer described in British Poultry Standards as it is believed they died out during the Second World War.

Special Management Requirements

Sicilian Buttercups are active birds, and their leg colour will also be helped, if they have access to large grassed runs. The original breeders (circa 1912 to 1920) claimed that Buttercups were very tame and docile, a characteristic that they thought came from many generations of having been kept in small flocks by Sicilian peasants.

The 1919 edition of *Wright's Book of Poultry* (revised by S.H. Lewer) gives suggestions for a double mating system to produce perfect markings on both sexes and also mentions the relationship between the quality of female feather markings on the back and wings in relation to a greater or lesser amount of breast spangling. There are no longer any large classes of Buttercups to win at even the major shows, so double mating is not necessary and the exact amount of breast spangling is not likely to be a deciding factor in awarding the prize cards in small classes. Apart from comb shape and general size and vigour, breeders should concentrate on selecting for correct 'ears of wheat' spangling on the back, tail and wings of females, and accept whatever amount of breast markings comes along. According to the advice given in Wright's, this may result in a too light general ground colour in the males bred from well-marked females. Some breeders might decide to accept this as a genetic inevitability, and effectively just do the 'pullet breeding' part of double mating. For those who prefer to show rich-coloured Gold Sicilian males, these were said to come from hens with markings more like those seen on Partridge Wyandottes than the 'Buttercup' pattern required on show Sicilian hens.

SICILIAN FLOWERBIRD

Some British Sicilian Buttercup breeders in the 1920s thought that the 'improvements' made by American breeders (Buttercups had come to the UK from America) had gone too far away from the original type in Sicily. They were particularly concerned about their weights, thinking them too large for a peasant village hen.

Fresh stock was imported direct from Sicily, and the breeders who agreed with this view formed a separate 'Sicilian Flowerbird Club' in 1922. This club does not seem to have lasted long, probably only until about 1928. The only record of a significant display of Flowerbirds was nine cockerels and sixteen pullets at the 1926 Crystal Palace Show.

During this brief period a standard was accepted by the PCGB. It specified weights of: 4 to 4½lb (1.8–2kg) for males and 3½ to 4lb (1.6–1.8kg) for females. Ear lobe colour was 'immaterial', they clearly did not want any arguments about this potential problem. The main plumage colour variety was the 'Mahogany', a dark variant of the Black-Red/partridge pattern. 'Spangled' was the other variety (as mahogany + white spots) but they were even rarer. The author has never even seen a picture of a Spangled Sicilian Flowerbird. The only coloured illustration of Mahogany Sicilian Flowerbirds from the decades when they existed was on one of the 'Players' cigarette cards issued in 1934. The artist was probably Ernest Wippell.

COVENEY WHITE/WHITE CORONET

Two breeders were involved with this idea, more or less a White Sicilian Buttercup, or a cup-combed White Leghorn, depending on your viewpoint. Mr W.J. Unwin of Cambridgeshire began the process, sometime during the 1914–18 war, when he started to make his 'White Coronets'. He crossed Sicilians (Buttercups? Flowerbirds?) with White Leghorns, in an attempt to make a more

Cup-Combed Breeds

compact-bodied version of White Leghorn that would also be less susceptible to frostbite in the comb than normal Leghorns. Mr Unwin was in the garden plant and seed business that went on to become 'Unwin's Seeds', a major UK brand in 2005.

In 1923, Stanley Street-Porter of Coveney Manor, near Ely, also Cambridgeshire, became involved in the project. He had enough wealth and influence to persuade the PCGB to accept a standard for the now renamed 'Coveney Whites' in 1924. Apart from the cup combs, this new breed was almost identical to smallish, utility strain White Leghorns.

They never attracted much interest, and had probably died out before the Second World War started. Similar birds would be fairly easy to re-create by crossing Sicilian Buttercups with white egg hybrids, but it is unlikely that they would be any more successful second time around.

AUGSBURGER

An article in 'Deutschen Kleinter-Züchters' (Reutlingen, 25 December 1956) on Augsburgers mentioned a book published in 1798 by Dr J.Ch. Gotthard in which 'Crown Cocks' were described in the Erfurt area. This was probably a chance strain, as the Augsburger breed was not made until the 1880s, and the city of Augsburg is a considerable distance south of Erfurt. As the article says, cup combs can be expected to appear in a flock that includes crested breeds. Appenzeller Spitzhaubens, Brabanters, Crève-Cœur, Houdans, Polands and Sultans all have horned or other forms of 'Duplex' combs, the gene responsible for cup combs.

Breeders in and around Augsburg, and others in the Rhineland, developed a more uniform type and colour. Herr J. Mayer is credited with producing the first uniform

Augsburger, pair. Artist: Kurt Zander

Cup-Combed Breeds

Augsburger, large male. Photo: Hans Schippers

Augsburger, large female. Photo: John Tarren

strain of Black Augsburgers by using La Flèche and 'Lamottas' (similar to Black Leghorns?). Without selective breeding, most 'Black' breeds have silver or gold neck hackles on most specimens. There were also White Augsburgers circa 1880–90. These died out, possibly during the 1914–18 war, and were not seen again until about 1990, a century after their first appearance.

Somehow Augsburgers survived through Germany's turbulent history to the late 1940s. They were being bred in both West and East Germany in the 1950s, and it was in the DDR (East) that Blue and Barred varieties were made. As far as is known, they were never used for large-scale egg production, but were kept by smallholders and hobbyists. The 1954 Augsburger Breed Club Show had an entry of sixty birds. Show entries appear to have been much lower in recent years.

Augsburger Bantam/ Zwerg-Augsberger

According to Wandelt and Wolters these were made during the 1930s, but this strain did not survive the Second World War. Otto Knöpfler started again in the 1950s by crossing large Augsburgers with Deutsche Zwerghühner (German Bantams) and Zwerg-Italiener (± Leghorn Bantams). He had some good enough to show by 1963. As far as the author is aware, they still exist, but in small numbers, almost all in Germany.

Description

General body size and shape is similar to other European Light Breeds, with cockerels being very meaty for their size and, like La Bresse, they make high quality, if small, table birds. Standard weights in the 1950s ranged from 1.5kg (3¼lb) pullets up to 2.5kg (5½lb) adult cocks. Later, by 1970, these had been increased by 500g (17½oz). Bantam weights are 900g (32oz) male, 800g (28oz) female.

Cup combs on Augsburgers are noticeably more upright than those of Sicilian Buttercups. The author does not know how many, or more likely how few, of each group hatched have combs of the correct standard shape. Their white ear lobes are of moderate size, so red in the lobe is more likely to be a problem than white in the face. Dark eyes are required, which if the photographs seen by the author are typical, suggest that dark pigmented faces are a common problem on pullets. Shanks and feet are slate-blue. Plumage is glossy greenish-black, which has been said to be difficult to achieve on females. In addition to the short-lived White variety, Barreds and Blues have also been created, at first in the mid-1950s in the former DDR.

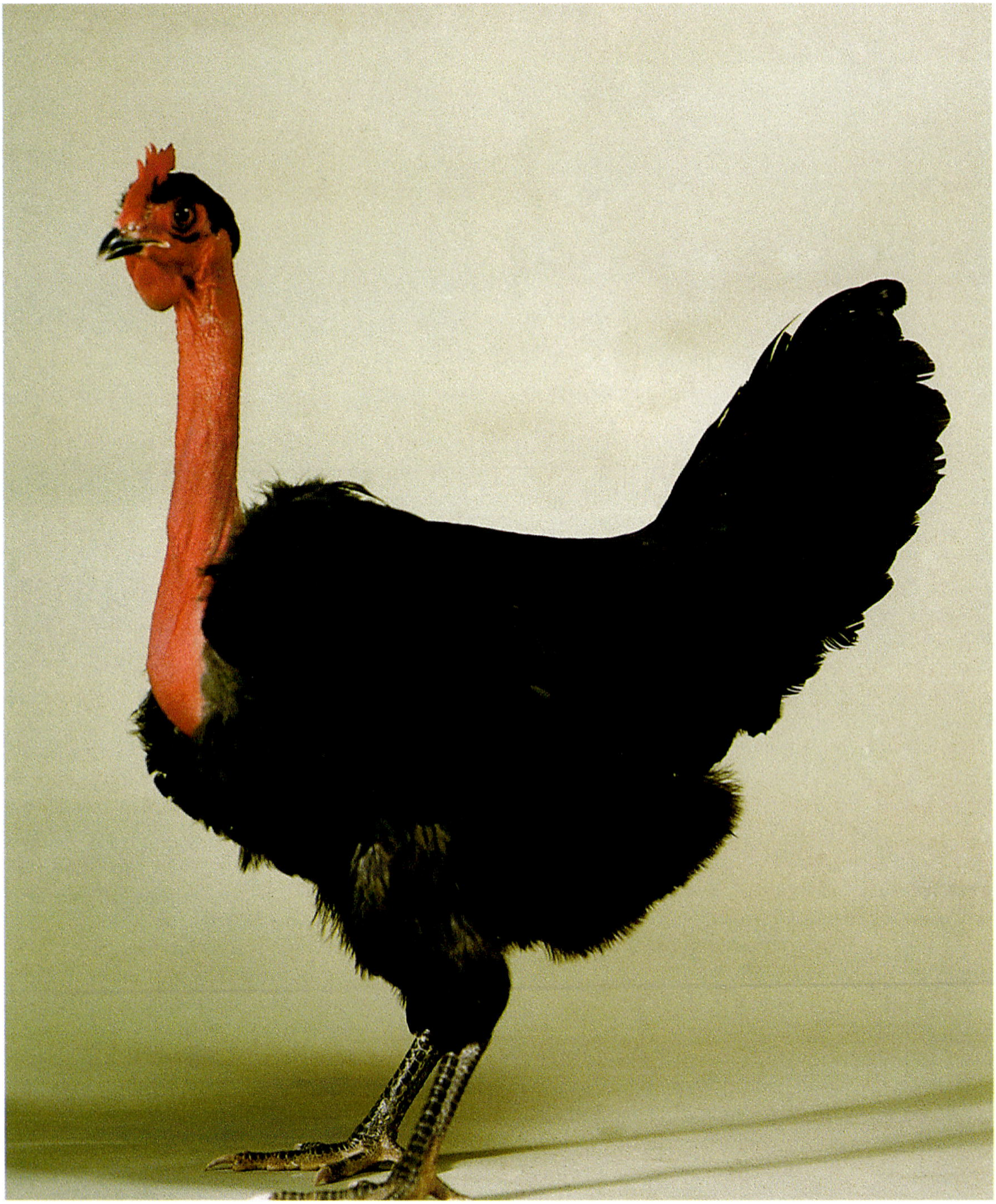

CHAPTER 8

Frizzled and Naked-Necked Breeds

This chapter would have included silkie-feathered breeds as well, but as Silkies and silkie-feathered Chabo/Japanese Bantams are too popular to be included in this book of rare breeds, they are not. Frizzle-feathered Polands and Chabo/Japanese are excluded for similar reasons, they are rare varieties of popular breeds.

This chapter therefore covers:

Annaberger Haubenstrupphuhn – Germany
Frizzle – British Standard form
Lyonnaise – France
Madagascan Naked-Necked Game – Madagascar and elsewhere in Indian Ocean area
Transylvanian Naked Neck – Romania, once part of Austro-Hungarian Empire.

ANNABERGER HAUBENSTRUPPHUHN

Georg Jonientz of Annaberg-Buchholz in the Erzgebirge region of eastern Germany made these, circa 1950. The two main breeds used were Appenzeller Spitzhaubens and Frizzles, possibly also Chabos, Polands and Sitkies. Annaberger Haubenstrupphuhner are very similar to Lyonnaise (*see below*). The author does not know how many still exist.

They are a small breed, with standard weights of only 1.5kg (3¼lb) for males, and 1.2kg (2½lb) for females, with a general body shape similar to Appenzeller Spitzhaubens, also with their horned comb and upright crest, although in this case composed of frizzled feathers. Annabergers are not bearded, having large wattles, white lobes and slate-blue shanks and feet. All plumage curls outwards. There are three colour varieties: Black, Black-Mottled and White.

FRIZZLE

British Poultry Standards include a breed called the 'Frizzle', with The Frizzle Society of Great Britain to support it. Frizzle-feathered Chabo/Japanese and Polands are covered by their respective breed clubs in the UK.

By contrast, the National Frizzle Club of America covers frizzled versions of any breed of chicken or bantam anyone chooses to make, plus Sebastopol Geese. American Frizzled Plymouth Rocks are the nearest equivalent to UK Frizzles. They also have Frizzled Polish and Japanese (Chabos) in the USA as in Europe. In September 1998 the National Frizzle Club of America listed the above, plus large and bantam Cochins, Rhode Island Reds, Leghorns and Naked Necks. They also have bantam only frizzled versions of White Cornish (Indian Game), Black Orpington and Modem Game (colours?). Australorps in Black, Blue and White are also mentioned, but it is not specified if they are large or bantam.

Frizzle-feathered chickens of various shapes, sizes and colours have existed for many centuries in many parts of the world, the

Frizzled and Naked-Necked Breeds

American White Frizzle Cochin, bantam male. Photo: John Tarren

details of which will probably never be known. Normal-feathered Chabos/Japanese were recorded at least as far back as 1600. Frizzle-feathered Chabos could have appeared at any time since then. Full-sized Frizzles have been reported from the Philippines, on several Indian Ocean islands and in many parts of Africa. Ulisse Aldrovandi described and illustrated Frizzles in 1600, birds owned by Pompilio Tagliaferro of Parma. This ancient book was translated into English by L.R. Lind and published by the University of Oklahoma, 1963.

Moving on to the mid-nineteenth century, and the beginning of poultry showing, Dr John Bennett described the 'Frizzled Fowl' recently imported into the USA in his Poultry Book published in 1850. Frizzles were then a specific breed in the USA, as in Britain today, not frizzled versions of other breeds as is now the case there. Dr Eben Wight of Boston and George C. Peirce and Stephen Osborn, both of Danvers, Massachusetts were Frizzle breeders mentioned. They had Blacks and Whites that appear from the illustration to have been like any other medium-weight chicken in other respects. Frizzles were entered at the first American poultry show held on 15–16 November 1849. Dr Bennett thought Frizzles originated in Japan, but it is more likely these came from West Africa, perhaps brought over on a slave ship. Mr G.T. Williams, British Frizzle Club Secretary mentioned (in 1923 *Poultry* yearbook) that the famous explorer, Dr David Livingstone, had recorded seeing Frizzles in West Africa.

There was a 'Frizzle' breed in the USA, as in the UK now, at least up to 1924. The author has not discovered when American rules changed to the current frizzled versions of other breeds. Only a few breeders kept them, but most of them were regular exhibitors, so they seemed to be more popular than they really were. John H. Robinson, in *Popular Breeds of Domestic Poultry* (1924) said that some birds appeared at twenty or more shows in a season. The leading American breeder of Frizzles in the early twentieth

White Frizzle, large male. Photo: John Tarren

White Frizzle, large female. Photo: John Tarren

century was T. Farrar Rackham of New Jersey who had an excellent strain of large Buff Frizzles of approximately Plymouth Rock type. Other breeders had Blacks and Whites.

Frizzles were also being bred in Germany, with some Frizzle Bantams among the entries of the first German National Show at Leipzig in 1893. They were said to have come from somewhere in the Austro-Hungarian Empire.

As in most other places, Frizzles were bred in very small numbers by lovers of the unusual in Britain. James Long mentioned (in *Poultry for Prizes and Profit*) that 'a titled lady fancier' entered 'the best specimens we remember' at the 1871 Birmingham Show. They were Whites. Interest was boosted by the formation of the Frizzle Club in 1912, the first Secretary being Miss Mullally of Co. Tipperary. Classes were arranged for Frizzles in the 1912 Crystal Palace Show, where four males and seven females appeared. 1st, M and 1st, 2nd F were won by Lady Alexander,

who may or may not have been the 'titled lady fancier' mentioned above. These were large Frizzles, and it was claimed that they were as active and productive as any other breed. Both single and rose combs were allowed, and the colour varieties listed were: White, Grey, Black, Red and Buff. At the same show, Frizzle Bantams were in the 'Variety Bantam Club' section, an organization that existed circa 1900–25. There were three classes for Frizzle Bantams: White M (four entries), White F (six), AOC M/F (seven). Most of the prizes went to Major G.T. Williams (the next Frizzle Club Secretary and Japanese Club Secretary) and Miss C.S. Entwisle, daughter of the famous W.F. Entwisle.

The 1914 *Feathered World Yearbook* included the first Frizzle Club Breed Standard, that still allowed both single and rose combs. Weights were 7–9lb (3–4kg) for males and 5–7lb (2.2–3kg) for females. Colours then listed were: White, Black, Blue, Red, Buff, Grey and Spangled. By 1920, as given in the 1921 *FWYB*, the Frizzle Standard only allows single combs. The Frizzle Club had been revived after the First World War by Major G.T. Williams, the new Secretary. The Major stressed how difficult it was to say where Frizzles had originated. He had seen them in 'Kashmir, India, Egypt, France, Belgium, etc'. This suggests he had led an exciting life. In 1922's *FWYB* there is a photo of a winning Spangled Frizzle hen, the property of Sir Claud Alexander (Bart). This shows that 'Spangled Frizzles' were of the Millefleur pattern, as Belgian Bantams.

Major Williams gave more details of Frizzles he had seen around Europe in the 1926 *FWYB* helps to give an idea of the distributuion of the breed. The Comtesse de Bylande and a nearby doctor in Belgium both bred a strain of Buff Frizzles that were rose-combed, crested, had five toes and some leg feathering. They were said to be halfway between large fowl and bantams in size. These details suggest some Silkies had been among the ancestors of this strain. Frizzles were also seen in Germany and the Somme area of France, surprising considering the devastation caused by the war. Frizzled Polands were seen in The Netherlands, but not up in the northern province of Friesland. Frizzles had often been called 'Frieslands' but, as the Major suspected, they had no connection with the area beyond a similar sounding name. He concluded by stressing that the (UK) Frizzle Club had set a standard for them, and variations like the Comtesse's strain would not be acceptable at British shows. The uniformity of British large Frizzles, compared to the variable range of shapes and sizes bred on mainland Europe, was confirmed in 1930 when fanciers from everywhere gathered at Crystal Palace for the World Poultry Congress. Frizzles were not popular enough to have classes at every show, so the Frizzle Club concentrated on the three 'classics', Birmingham, Crystal Palace and the London Dairy Show, and eight other selected venues. Major Williams and Sir Claud Alexander remained the two main exhibitors of large Frizzles in the UK up to 1939. The original Frizzle Club was not revived after the Second World War; the new Frizzle Society of Great Britain was not formed until 1978, the breed having spent a decade (since 1969) in the Rare Poultry Society.

Frizzle Bantam

Frizzle-feathered Chabo/Japanese Bantams probably date back several centuries, and there have been many other frizzle-feathered birds bigger than Chabos, but smaller than 'Large Frizzles'. The present day type and style of Frizzle Bantam were probably established about 1880. A Mr Edward Morgan was mentioned (in the 1934 *FWYB*) as being one of the pioneer breeders. He said that back in 1880 Frizzle Bantams were a very variable lot. His much more uniform strain of Buff Frizzle Bantams was made from the best Buff male he could obtain in the UK with two 'Nankin' bantam hens imported from Yokohama on the S.S. *Flintshire*.

Description

Their bodies are short and broad, with full-rounded breasts. Frizzle breeders regularly get some smooth feathered chicks. Some of them, usually pullets, are kept for future breeding because constant Frizzle to Frizzle mating gives a dramatic loss of feather quality, eventually leading to some pathetic-looking half-naked creatures. The smooth-feathered birds are roughly similar to other medium/heavy breeds, utility type Rhodes or Rocks perhaps. Standard weights for large Frizzles range from 2.25kg (5lb) pullets up to 3.6kg (8lb) adult cocks. The equivalent bantam weights are 570g (20oz) and 790g (28oz). Quality and even curl of plumage is the most important feature. They have straight single combs, red eyes and red ear lobes. Most all-round judges expect yellow beak, shanks and feet on all colour varieties, but the Frizzle Society of GB Standard only requires this on Buffs, Columbians, Piles, Reds and Whites. White, willow, blue and black shanks and feet are allowed on other colours. Relatively few shows are judged by Frizzle specialists how-

Frizzled and Naked-Necked Breeds

Black Mottled Frizzles, bantam trio. Photo: John Tarren

White Frizzles, bantam trio. Photo: John Tarren

ever, so breeders will still find yellow-legged birds winning most of the prizes.

LYONNAISE

Similar to Annaberger Haubenstrupphuhner, this French breed is also frizzled and crested. They were bred from Black Hamburghs and Frizzled White-Crested Black Polands. The main colour variety is Black, and weights range from 1.5kg (3¼lb) pullets up to 2.5kg (5½lb) adult cocks.

MADAGASCAN NAKED GAME

Virtually naked Malay or Shamo-type fighting cocks have been bred for centuries all around the Indian Ocean. As might be expected, they are very large, very tall, and have typical Oriental Game type headpoints. All skin is bright red as they are only fully feathered on wings and tail. They probably originated in South-East Asia, Indonesia, Malaysia, Vietnam, and so on and could have been taken to Madagascar at any time over the last 2,000 years. Although Madagascar is off the coast of Mozambique, East Africa, much of its population and culture is more South-East Asian than African. Madagascar had contacts with the Portugese back in the sixteenth century, eventually becoming a French colony in the nineteenth century, when they were noticed by French fanciers, and named 'Denude de Malgache'.

Virtually naked fighting chickens have a distinct advantage in the heat of the tropics, be it Madagascar or Vietnam, where there are very similar birds. Recent era Vietnamese cockers have used secret creams and lotions to harden the skin of their birds for combat, and similar methods may have been used for centuries in Madagascar as well. Those few Madagascan Naked-Necked Game imported by European fanciers have been kept purely as curiosities from a far-off land. They would probably be comfortable enough in the south of France, but anyone attempting to keep them in northern Europe would have enough trouble just keeping them alive and in good condition. Heated houses are probably essential.

TRANSYLVANIAN NAKED NECK

Standard bred Naked Necks, as bred in Europe in limited numbers since the nineteenth century, were probably bred from crossings between imported Madagascan Naked-Necked Game and local farmyard chickens. More or less naked-necked chickens have been seen all around the Mediterranean area and up into the Balkans.

Transylvania, the nominal place of origin of the standard Naked Neck breed, has been part of Romania since 1918. From 1700 until 1918 it was part of the old Austro-Hungarian Empire, and before that Transylvania was part of the Ottoman Turkish Empire. The author has no further details, but Naked Necks are said to have been mentioned in an Austrian book published in 1701. In the show world, Naked Necks have been much more popular in Austria and Germany than anywhere else, and it was these breeders who selected for completely naked necks and specific comb and body shapes.

Naked Necks were first seen at a European poultry show in 1875, when Herr C.E. Weber entered some at a show in Vienna. Progress was rapid, including the development of a bantam version that is detailed below. There was a wonderful display of 110 large Naked Necks at the Leipzig Show in 1907. A breed club, the Nackthalszüchter-Verein was formed by 1911. The author has evidence of this in the form of a picture postcard published in Leipzig that year by the club. Artwork was by Max Loose, and illustrates a Black male, Cuckoo female and White female. Naked Necks have never been among the really popular breeds, but have practical qualities to retain a substantial number of breeders. Not only do they have naked necks, but they have far fewer body feathers than other breeds. Surplus cockerels are easily and quickly hand-plucked, a significant

Postcard published by the Nackthalszüchter-Verein, 1911. Artist: Max Loose

advantage when a lot of cockerels have to be done.

Although not the most popular breed, by 1955 there was an estimated 600 Naked-Neck breeders, large and bantam, over all of the then separate East and West Germany. Herr B. Noack wrote 'Das Nackthalshuhn' (1958), almost certainly the only specialist book ever written about this breed. There are not as many Naked-Neck breeders in Germany now (2005), but there are still quite a few.

64th German National Show, Dortmund, 1982
  – *Large*, Black-13, White-2, Blue-3.
  – *Bantam*, Black-9, White-3, Blue-3.
104th Hanover Young Bird Show, 1985
  – *Large*, Black-35, Red-4, Buff-3, Blue-5.
  – *Bantam*, Black-11, White-7, Barred-5.

Naked-Neck Bantam

As can be seen from the figures above, Naked Neck Bantams are not now a very popular breed, but are kept by several enthusiastic fanciers. The first examples were shown by Karl Huth at the 1898 German National Show at Frankfurt. Otto Marhold, active circa 1905–15, was the next breeder the author has details of. He made Black, White and Black-Red/Partridge Naked Neck Bantams by crossing under-sized large Naked Necks of each variety with suitable bantams, the details of which are not recorded. Any crossbred bantam would do, as long as they were roughly the required size, shape and colour. It is not known if any of these early strains survived the First World War, but two breeders in the early 1920s made new strains. Richard Muller of Meuselwitz used Black Rosecomb Bantams to help make some Black Naked Neck Bantams and Michael

169

Frizzled and Naked-Necked Breeds

Black Translyvanian Naked Neck, large female. Photo: John Tarren

Black Translyvanian Naked Neck, bantam female. Photo: John Tarren

Drexter of Munich made Black-Red/Partridge Naked-Neck bantams from the same colour of Deutsche Zwerghuhner (German Bantams). Both breeders would obviously have needed some Naked Necks as well.

They had to be made from scratch again in the 1950s by a new generation of fanciers. Kurt Hünlich of Neusalza-Spremberg had made Red Naked Neck Bantams by 1956. He had started from Rhode Island Red Bantam F × (undersized?) large Black Naked Neck M mating. All colour varieties had been remade in both East and West Germany by the late 1970s.

Naked Necks in the UK

There have never been many Naked Necks in the UK, certainly nowhere near as many as in Germany. Mr John C. Fraser was the first importer and exhibitor in England in 1874. His birds came from Transylvania, which that as said above, was then a province of the Austro-Hungarian Empire. The next mentioned were shown by Lord Deerhurst at the 1900 London Dairy Show. Apart from the

Black Translyvanian Naked Neck, bantam male. Photo: John Tarren

name 'Transylvanian Naked Necks', obviously coming from the origin of Mr Fraser's birds, they were sometimes called 'Bosnians' (on cigarette cards issued by Ogden's in 1916

White Translyvanian Naked Neck, bantam male. Photo: John Tarren

Cuckoo Transylvanian Naked Neck, bantam male. Photo: John Tarren

Blue Transylvanian Naked Neck, bantam female. Photo: John Tarren. Although called 'Blue Naked Necks', they are actually Leavenders.

for example), which suggests there may have been further importations from elsewhere in eastern Europe. Although most British poultry books published from 1914 to 1950 mention Naked Necks as one of the strangest breeds, there are very few details of anyone actually keeping them here. Apart from their oddity, many British people were probably put off them, especially during 1914–18, because of their Austro-Hungarian connections. No products 'Made in Germany', or by its allies, were going to be top sellers then.

The 1934 *Feathered World Yearbook* included an article on 'Churkeys', one of the names sometimes given to Naked Necks, another in similar vein being Turkens'. Some people (very stupid people?) thought they were crosses between chickens and turkeys. A group of breeders had entered a display of them in the non-competitive class at the 1933 Crystal Palace Show. The bird illustrated was a rosecombed White cockerel with a large ruff of feathers on the front of its neck, a major show fault now. It was owned by Mr James Hayden of Hartley Witney, Hampshire, and to be fair were probably the best specimens available at the time.

Only poor quality Naked Necks, with a lot of feathers on the front of their necks, were seen in Britain until the late 1970s, when hatching eggs were brought over from Germany. The bantam version came over for the first time around then as well. T.V. Richards from Redruth, Cornwall entered a

Black cockerel at the 1980 National Show. It attracted a lot of interest and admiration from the Rare Breeds judge (me!) and everyone else interested in the unusual. Despite good-quality Naked Necks being available, neither size version has attracted many breeders in the UK; probably less than twenty serious fanciers overall by 2005. At least one large Naked Neck breeder was planning to improve size and vigour by crossing exhibition stock with naked necked broilers. It will take several years of selective breeding to fully restore show bird type and colours, whilst retaining the increased size from the commercial meat birds. Naked Necks are covered by the Rare Poultry Society in the UK.

Naked Necks in the USA

As in Britain, only a few Naked Necks were bred as curiosities in the USA during the whole of the twentieth century. 'Churkeys' and 'Turkens' were also sold to the gullible as chicken/turkey hybrids.

Naked Neck Bantams were imported by Bernard L. Kellog of Viroqua, Wisconsin from Germany in the 1960s. Large and bantam Naked Necks were standardized by the APA in 1965 in four colours: Black, Buff, Red and White. Mr Kellog was still advertising them in the 1981 *ABA Yearbook*, possibly later (the author only has the 1980 and 1981 editions). He was keeping '2 colors', but did not specify which.

The December 1988 edition of *Poultry Press* mentions the 'National Naked Neck Breeders Society' as a new club. Show results in the same magazine mention large Naked Necks in Barred, Black, Crele, Red and White with just one mention of Black bantams. The NNNBS was still active in 1998. The author had no further details available.

Description

Apart from their naked necks they have much reduced body feathering as well. The main feather growth areas ('tracts') are smaller than those of other breeds and the areas of skin in between are usually completely bare on Naked necks, compared to the down covering of the same areas on normal chickens. This no doubt accounts for their popularity over the centuries in those parts of the world warm enough for them to be comfortable. Few people would claim they are the prettiest breed, even the enthusiasts. Interesting yes, beautiful no. But plucking chickens for the table is a chore no one particularly enjoys, and it doesn't take long to pluck a Naked Neck. They all have a 'cap' of feathers on the head that should taper to a neat point at the rear. The neck should be completely clear of feathers, although many have few feathers (often removed for showing) halfway down the front. All the bare skin should be bright red, and this usually extends down to the upper half of the crop. Although rose combs have been common in the past, upright single combs are now expected.

They are a medium/heavy breed. When standing normally, the back line should slope down slightly. Their tails are fairly long and are held a few degrees above the horizontal. The 1997 British Poultry Standards specify 'alert, upright and bold' carriage and the tail to be at an angle of 45 degrees. Birds like this would not do very well at a large German show, the main centre of Naked Neck showing, so the author hopes that these details are changed in future editions.

Standard weights are lower in Germany than in the UK and USA. The large version in Germany ranges from 2kg (4½lb) pullets up to 3kg (6½lb) adult cocks, with the British and American weights being about 500g (18oz) heavier for each category. Naked Neck bantams are sometimes too small, which is unusual among bantam breeds nowadays. Expect them to range from 700g (25oz) pullets up to 950g (33oz) adult cocks.

Although Black-Red/Partridge and Pile Naked Necks are seen, most are self colours: Black, Blue, Buff, Cuckoo, Red and White. Blues should be clear coloured (not laced) if possible, and Reds are as Rhode Island Reds.

Special Management Requirements

Free range is essential for Naked Necks. Sun, wind and rain will help keep the exposed skin bright red. If kept inside for long periods the skin will fade to pallid pink. This creates a dilemma for breeders of Blues, Buffs and Whites that would normally be kept in the shade to preserve plumage colour. During the moulting season they often start feather pecking, mainly the new (full of blood) feathers growing at the crop. It may be necessary to pen a lot of birds individually (in show cages?) for a few weeks.

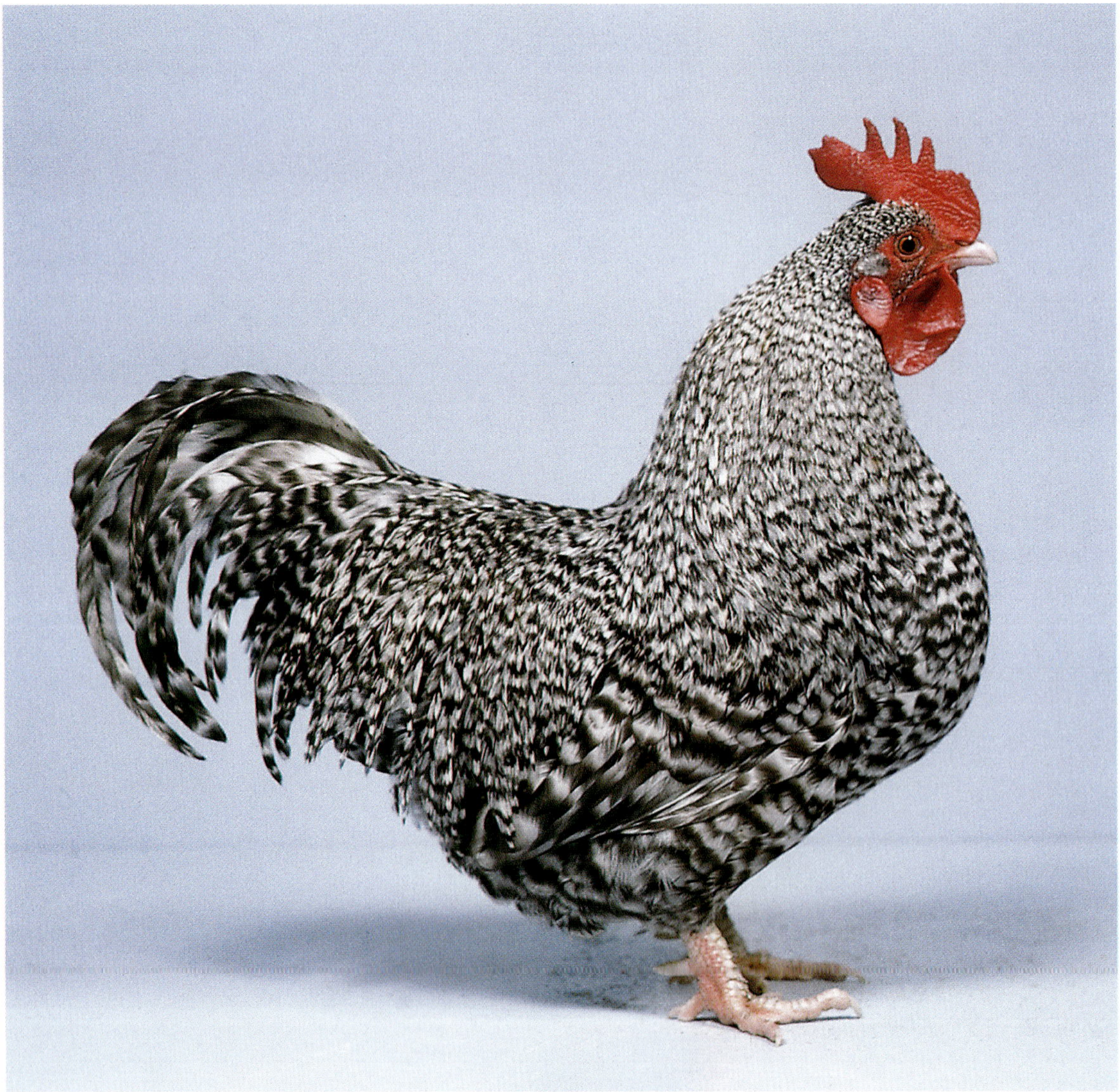

CHAPTER 9

Short-Legged Breeds

Four of the five breeds in this chapter, Courtes Pattes, Krüper, Luttehøne and the Scots Dumpy are very similar, being essentially the French, German, Danish and Scottish versions of the same thing. The other breed, the Jitokko, is from Japan and therefore not closely related.

Short-legged breeds have been known in Europe at least since the Renaissance period, and at that time were obviously not bred to any standard. Their collective ancient history is not readily available, and would probably not be relevent anyway. Not very much of the history of the other breeds is available either, so most of this chapter is about the Scots Dumpy.

COURTES PATTES

These are from the area around Le Mans in the Sarthe Département of France. Like all of these short-legged breeds, they are excellent broodies and mothers. They were mainly bred by small-scale poultry keepers. Gourmets appreciated their fine qualities as table birds, but Courtes Pattes were too small to be saleable at the normal poultry markets. Their current standard weights range from 1.5kg (3¼lb) pullets up to 2.25kg (5lb) adult cocks, and these weights are an increase on their size in the late nineteenth century.

Blacks are the main colour variety, for many years the only colour; Barreds, Black-mottles and Whites were added during the twentieth century. All have red eyes and white ear lobes. Shanks and feet are white on Barreds and Whites, dark slate on Blacks and light slate on Black-Mottles.

JITOKKO

These originated on the southern Japanese island of Kyushu. It is thought that they became a fixed breed sometime during Japan's self-imposed period of isolation from the rest of the world (1600–1868). As a result of this being a closed society, few details are available. One suggestion is that the name 'Jitokko' is derived from a type of local official. It may be that the cocks, when strutting about on their short little legs, reminded the downtrodden peasants of some pompous official who was not making their lives any easier. Thus to name their chickens after the hated officials was a tiny streak of rebellion.

Jitokko are larger (1.9kg (4lb) pullets to 3kg (6½lb) adult cocks) than the European breeds in this chapter, and they have different head character and tail carriage. This is no doubt a result of crossing over the centuries with other local breeds, such as Satsumadori.

Although single and walnut combs are also occasionally seen, most Jitokko have pea combs. Beside and behind the comb there is a small crest, and they also have a beard. Tails are of medium length, fanned and sweep out and down from the back line to the ground. Shanks are short and thick. The main colour variety is Black-Red males with brown (with

Short-Legged Breeds

irregular black peppering) females. There are also Blacks and Whites.

KRÜPER/GERMAN CREEPER

Edward Brown described Krüpers in his 1905 book *Races of Domestic Poultry*, which suggests that they must have been a recognized breed for several decades before that time. He noted that, like the French Courtes Pattes, they were fairly rare and mainly bred by amateur poultry keepers. Like all their relations they are naturally tame, good mothers, fair layers and good quality, if not very large, table birds. Krüpers were mainly bred in the German state of Nordrhein-Westfalen.

Zwerg-Krüper/German Creeper Bantam

The miniature version of this breed was first made in the 1920s, but died out during the Second World War. They were made again about 1990. Breeds used to make them were probably the obvious undersized large Krüpers, Chabo (Japanese Bantams) and Deutsche Zwerghühner (German Bantams).

Description

General size, shape and colours are similar to their European relations. Combs are single and smaller than those of Courtes Pattes. Krüpers have circular, white ear lobes. Eyes are dark red, at least on the darker colour varieties. The colour varieties are Black, White, Cuckoo, Black-Red/Partridge, Silver Duckwing, 'Schwarz-Gelbgedobbelt' (as Bergischer Kräher) and 'Schwarz-Weißgedobbelt' (as Bergischer Schlotterkamm). Standard weights for the large fowl range from 1.5kg (3¼lb) pullets up to 2.25kg (5lb) adult cocks, bantam males are 900g (31oz) and bantam females are 800g (28oz).

LUTTEHØNE

This is the Danish equivalent to Courtes Pattes, Krüpers and Scots Dumpies. Most of the short-legged breeds are related to a normal leg length breed in their home area, as suggested above with Krüpers and Bergischer Schlotterkamm, and below with Scots Dumpies and Scots Greys. Luttehøne are related to Dansk Landhøne, and were first officially recognized as a breed in 1901.

Dwaerg-Luttehøne

The bantam version of Luttehøne is a recent creation, circa 1975.

Description

Body shape, size and so on are much the same as the others. Standard colours are Black, White and Black-Red/Partridge, the last being the most popular as it is with the longer- legged Dansk Landhøne.

SCOTS DUMPY

Fortunately for the author, considerably more information is available on Scots Dumpies than the other short-legged breeds; otherwise this would have been an embarrassingly short chapter.

Some Victorian poultry books said that they have been bred in Scotland for centuries, but do not give detailed evidence. They have at least existed long enough to have been given several local or nicknames: Bakies, Creepies, Daidlies, Go-laighs, Hoodlies, Kitties and Stumpies. Since Scots Dumpies became established as an exhibition breed, the two main colour varieties have been Blacks and Cuckoos ('Greys'), but back when the shows were just starting, about 1850, and presumably for a century or so before then, the main colours were Silver-Hackled Blacks (as Norfolk Greys), Gold-Hackled Blacks and Speckled Reds (as Speckled Sussex).

Mr Fairlie of Cheveley Park, just south of Newmarket, Suffolk was the first person known to have exhibited them, at 'The Bazaar', Baker Street, London at Christmas 1852. There may not have been any poultry shows in Scotland by this date, so we should not be surprised that a Scottish breeder did not enter some earlier. The famous poultry expert, Harrison Weir wrote about Mr

Scots Dumpies and Scots Greys, pairs. Artist: A.F. Lydon. Originally a free gift with the Feathered World *magazine, 3 December 1920. As with many* Feathered World *prints, the first batch was printed with the issue date, but readers could send for them many years later. Stock was topped up by later printings, without the date.*

Fairlie's flock fifty years later in 1902: 'The flock consisted of some twenty-five or thirty, and all bore the stamp of being a separate and distinct breed from any other European fowls.' From Weir's drawing and description, they were much the same as Scots Dumpies today. 'In colour they were red and black, or yellow, grey and black, splashed with white, in some instances quite "gayly" so.' This is not a very clear description, but it can be assumed that the former were Gold-Hackled Blacks with white mottling, and the latter were the same plus the barring gene, to make a modified Crele, also plus white mottling.

Early editions of Lewis Wright's *Book of Poultry* said that Scots Dumpies declined in numbers in the 1870s, possibly because the rise in interest in poultry generally tended to promote those breeds that were receiving most attention at the shows. Scots Dumpies were not being exhibited much then. No doubt there were enough Scots Dumpies still being kept by smallholders in their main homeland, a band across central Scotland from the Isle of Arran in the west to Aberdeen on the east coast.

Interest in Scots Dumpies among exhibitors began to pick up in about 1910. A class was provided for them at the 1911 Crystal Palace Show that drew fifteen entries. A 'Scots Dumpie Club' was formed in 1912. The new club provided two classes at Crystal Palace in 1912 and held their first club show at Kilmarnock on 1–2 January 1913 that attracted twelve males and thirteen females. Most of the winners were Cuckoos and Blacks, which would set the preference among exhibitors and judges up to the present. Rapid progress was made during 1913, as by the end of the year the club had about thirty members, with another seventy known breeders whom they were attempting to recruit. The Club Secretary, James W. Brown said that the breed was nearly extinct a few years before the club was formed. Their second club show, at Airdrie on 29 November 1913, had four classes: nine cocks, fourteen hens, twelve cockerels and eighteen pullets (total fifty-three). Mr Brown was judging, and said that the two main faults that needed to be eradicated were white ear lobes (that should be red) and high tail carriage (that should be fairly low, to emphasize the general long, low, boat-like appearance required). Again, most of the winners were Cuckoos ('Greys') and Blacks.

Black Scots Dumpy, large male. Photo: John Tarren

Black Scots Dumpy, bantam female. Photo: John Tarren

After the First World War, the Scots Dumpy Club had a new Secretary, A.S. Paxton, and an excellent show at Edinburgh, 5–7 January 1921, where a total of sixty-four Dumpies were entered in eight classes, including some Scots Dumpy Bantams (*see below*). This was probably the high-water mark, as the following show (2–3 January 1922) was down to forty-four birds in seven classes. The decline continued; there were only twenty-four birds from four exhibitors on 1–2 January 1924 at Paisley. It was pointed out that the person entering the highest number of birds was Mr Major, better known for his Dorkings. He had travelled 500 miles up from Buckinghamshire for the show, so putting the local breeders around Glasgow to shame. Club Secretary, J.W. Brown lamented (in the 1931 *FWYB*) that very few of the breeders on the Isle of Arran, a traditional stronghold of Dumpies, still travelled any distance to show their birds. The author believes that the Scots Dumpy Club, like many other breed clubs, was suspended for the Second World War, and was not revived when peace returned.

In 1969, the Rare Poultry Society was formed to support fanciers of those breeds that did not have an active breed club in the UK. Dumpies had just about survived, Jim Smith of Terregles near Dumfries being one long-time supporter who kept them (and Scots Greys) going through the depressing decades when no one seemed interested in preserving old breeds. The few inbred survivors in the UK were boosted by the importation of some hatching eggs in 1973 from Mrs Violet Carnegie in Kenya. If any of the other European short-legged breeds were imported for emergency crossing, their import was not made public. It is possible that even if anyone had thought of doing so (and there was very little contact between British and Continental breeders then) they may have considered these breeds, all smaller than Scots Dumpies, not much of a help. The Scots Dumpy Club was re-formed in 1993.

Scots Dumpy Bantam

Work started on the production of Scots Dumpy Bantams about the time that the first Scots Dumpy Club was formed in 1912. Mr J. Craig of Dreghom, Ayrshire (south-west of Glasgow) was the informal leader of the project. Others joining in included Mr J. Lindsay of Arrochar and James Garrow of Loanhead, Midlothian. It is believed that they used the smallest possible large Scots Dumpies and Scots Grey Bantams to hasten the miniaturization process. It is not known if anyone tried Chabo (Japanese) Bantams. This breed has the obvious short legs and bantam size required, but brings the potential problems of yellow legs and high tail carriage.

By the early 1920s they were of excellent quality, often better than the large fowl, but never attracted more than a very limited number of enthusiastic breeders. As with the

Cuckoo Scots Dumpy, bantam male. Photo: John Tarren

Cuckoo Scots Dumpy, bantam female. Photo: John Tarren

large fowl, some of the breeders have also been Dorking fanciers, another breed with a long, low general outline. As a comparatively recent creation (not going back into unrecorded history, like large Dumpies) the bantam version of the breed have only been seen in Black and Cuckoo ('Grey'), without the range of unusual or ancient varieties.

Description
Scots Dumpies are roughly similar to the other European short-legged breeds in this chapter, but are larger and with slightly more profuse plumage. This should not be overdone, as was evident in the 1920s and 1930s, when some Dumpies in the shows were suspected of being crosses with Orpingtons. This increased the apparent body size, but rather too much, and at the same time reduced tail length, thus losing the essential long, low appearance, often said to be reminiscent of a rowing boat. Standard weights are: large male 3.2kg (7lb), large female 2.7kg (6lb); bantam male 800g (28oz), bantam female 675g (24oz). Eyes and lobes are red; beak, shanks and feet, are black or slate on Blacks, white with black spots on Cuckoos, white on other colours.

'No fixed plumage colour' is specified in British Poultry Standards for Scots Dumpies, but Blacks and Cuckoos are the two main varieties. Although the Standards, and many Scots Dumpy enthusiasts, may not worry about details of plumage colour (concentrating instead on body shape), most general judges will tend to give top awards to 'conventionally coloured' birds. Thus most winning Cuckoos will be evenly coloured all over, so breeders would be well advised to avoid light-hackled males. Whites can be considered the third 'traditional' colour, and going on from there, Silver Grey and Dark, as Dorkings, have regularly appeared as well. This is no doubt a consequence of many years of both breeds being kept by the same people. The other colours in British Poultry Standards are 'brown, gold and silver'.

Special Management Requirements
Scots Dumpies have their very short legs because they carry a single copy of the leg length-reducing 'Creeper Gene'. This is a 'semi-lethal' gene, which means that embryos with two copies of this gene do not hatch. The best short-legged exhibition males may not be very fertile because they are physically unable to mount their hens. Thus many fanciers solve both problems by mating their best short-legged hens with rather long-legged males that probably do not carry the creeper gene at all, but because of other modifying genes will still be a lot shorter in the leg than Scots Greys or other 'normal' breeds.

Housing, perches and runs should all be designed with their limited mobility in mind.

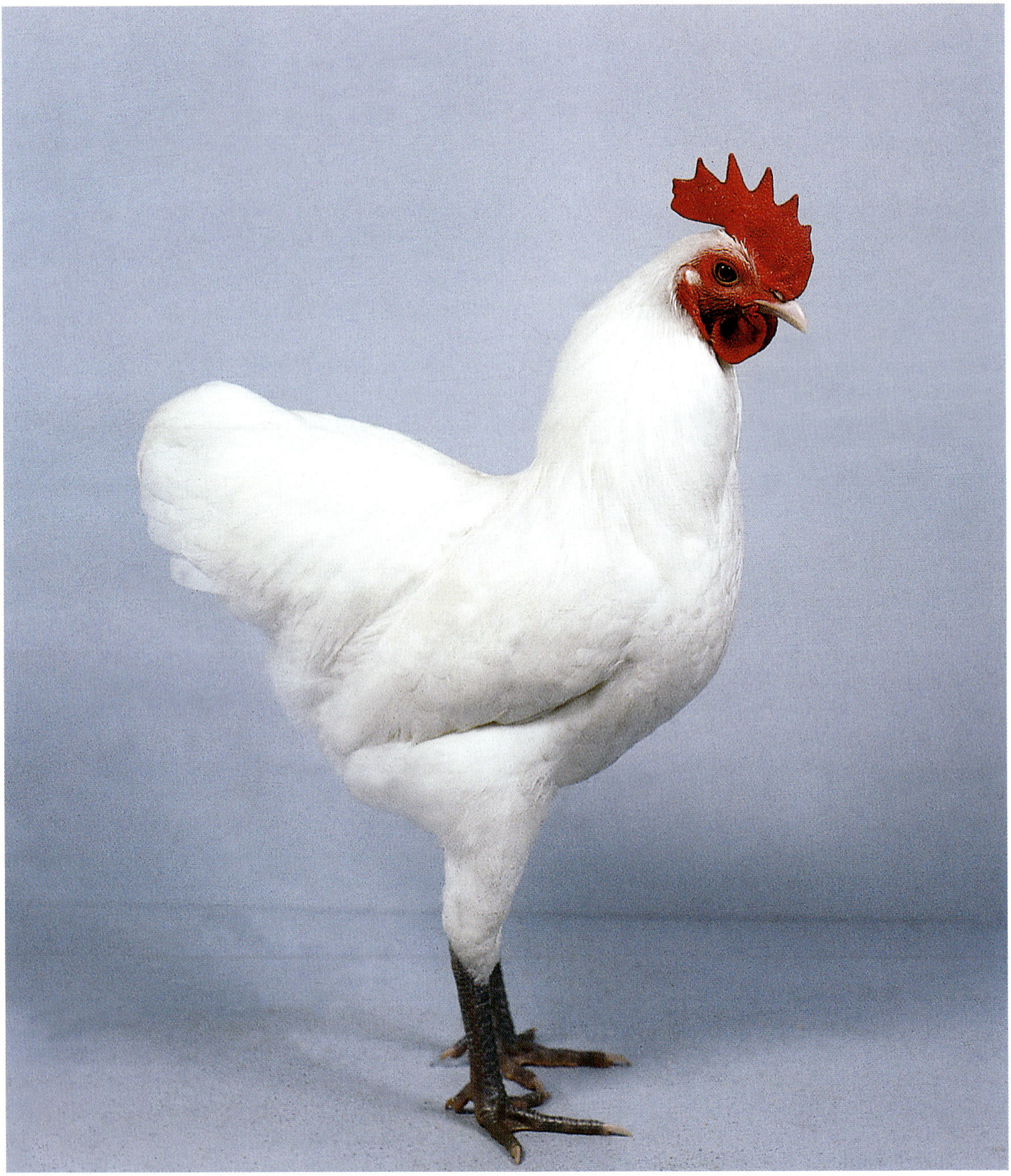

CHAPTER 10

Langshans

When Langshans, originally from China, first became known to poultry keepers in the Western world, some doubted whether they were different enough from the Black Cochins/Shanghaes that they already knew, to qualify as a distinct breed. Passions were raised and arguments raged in the letters pages of the poultry magazines of the late Victorian and Edwardian eras. Then the Langshan breeders were divided among themselves, resulting in Britain in two types of Langshan, 'Croad' and 'Modern' (or 'Society'-type).

American breeders remained aloof from all this, resulting in American Standard Langshans that are different again. Sheer distance resulted in yet another type, the 'Chinese Langshan' of Australia. In Germany and the then Austro-Hungarian Empire they had yet another ideal, resulting in 'Deutsches Langschans'. Miniature versions of all of the above have been produced.

This chapter covers:

Langshans – The first imports from China
Croad Langshans – British Standard
Modern Langshans, or 'Society-Type
    Langshans' – British Standard
Langshans – US Standard
Chinese Langshans – Australian Standard
Deutsches Langschans – Germany,
    Austria, Hungary

LANGSHANS – THE FIRST IMPORTS

Major F.T. Croad of The Manor House, Durrington, near Worthing, Sussex is usually credited with being the first importer and exhibitor of Langshans, showing three birds in the AOV class at the 1872 Crystal Palace Show. His birds may not have been the first, as Mr P.H. Bayliss, writing in the 1927 *Poultry Club of Great Britain Yearbook*, recalled reading in a November 1869 issue of *The Cottage Gardener* about a trio of Black Langshans that had been entered in a show at Edinburgh by a Mr Alexander.

Black Cochins that were first known as Black Shanghaes, had been known in Europe since the 1850s. By the 1870s British exhibitors had already started to change them from the original type, with light leg and foot feathering (like Croad Langshans today) to a heavier feathered exhibition type that would eventually (1900?) become the Black Cochins we see today. There had been numerous shipments of the large, more or less feathery, chickens from northern China to ports in Europe and America as gradually more and more regions of China were opened up, sometimes by force, for trade with the outside world. Many of these shipments were sold under colourful names, either by the seamen who were selling them or the poultry breeders who bought them to breed from and sell on. Some of the names given may have referred to the actual place of origin of a

Langshans

Black Langshans, circa 1890. Artist: J.W. Ludlow

shipment, others may have been place names that just sounded exotic. The British reached the Langshan area about 1862, so it is quite possible that Mr Alexander had the same stock as Major Croad.

Many poultry judges and experts were less than impressed by each importation being given a different name. They had to accept the name 'Cochin', which had been adopted by the rural community from earlier importations (Cochin-China was the old name for southern Vietnam), despite the fact that the name 'Shanghai' was more accurate for most of the birds imported after 1850. To them, Langshans were just another local variant of Black Cochins, and it was questionable whether they deserved a separate breed name.

Major Croad died a few years later, but his niece, Miss A.C. Croad had taken up the breed with enthusiasm. She was vehement in her defence of the Langshan as a distinct breed. She wrote and published *The Langshan Controversy* (1879), a 28-page collection of her correspondence to numerous poultry magazines. This was followed in 1889 by a more substantial book (122 pages), *The Langshan*.

Major and Miss Croad probably believed that the Langshan fowls they had recently imported had existed exactly as they were since the day God made them. Charles Darwin's *Origin of Species* (1859) and *The Variation of Animals and Plants Under Domestication* (1868), were still very controversial and had been widely condemned by the church. Victorian army officers were seldom radical thinkers, so it is likely that Major Croad and his niece followed the church's ideas rather than Darwin's.

Lewis Wright was one of the experts who fell out with Miss Croad. She had asserted that Langshans did not suffer from 'elephantiasis' (scaly leg mite infestation), and that this was proof that they were not related to Cochins that were well known for their susceptibility to this condition. Wright had seen Langshans with scaly leg at the 1878 Bath and West Show, held that year at Oxford (this was long before the present permanent site at Shepton Mallet), and was quick to point this out to Miss Croad. Wright was supported by famous poultry artist J.W. Ludlow when Miss Croad claimed that Langshans always had far glossier black plumage than any other breed, and that it was never spoiled by any red or yellow neck feathers. Both gentlemen had indeed seen red and yellow feathers on Langshan males, and they had seen many Black Hamburghs, Black Malays and others that were just as glossy as any Langshan.

A preliminary meeting of leading breeders was held on 18 August 1884 at 1 Bennett Street, St James, London established The Langshan Society. Miss A.C. Croad was one of those who attended, along with John Freeman, Harrison Weir, Reginald Herbert, G. Carvill and C.J. Barrett, the last named becoming Honorary Secretary. Major Croad also joined during the next year, there being a total of twenty-three members by the time of the next annual meeting at the 1885 Crystal Palace Show. At this meeting, Captain H.D. Terry took over as Honorary Secretary, and Reginald Herbert became President. This may have been the first hint of further troubles ahead, as one might have expected that either Major or Miss Croad would have been given that honour. There were further changes in the officers of the Society up to 1890 and, also in 1890, Miss Croad left it, soon to form a separate 'Pure Langshan Club', rapidly changed to 'Croad Langshan Club' after objections from the original Langshan Society. She was probably upset, among other things, by a motion at the 1887 AGM 'that special attention was drawn to the fact that feather should be as tight as possible'. This was clearly the beginning of the taller, tighter feathered, breed that would eventually become the Modern Langshan on one side, and a Croad Langshan that tended towards the compact and more profusely feathered end of the range of shapes that had been seen since 1972. For a few years, up to about 1905 perhaps, the differences between the two had not yet been made permanent.

In this early, all together as just 'Langshans' phase, the Society held some successful specialist shows, with entries of: 139 in 1887, 138 in 1888, rising to a magnificent 200 in 1898 and 199 in 1899. The 1898 Show was held at Lambeth Baths, London, where the Langshan Society members were the guests of the London and Suburban Poultry Society. Class entries were: twenty-two cocks, twenty-one hens, fifty cockerels, fifty-five pullets, thirty-three selling and nineteen entries in two 'local' classes. Remember, they were all large fowl, and all black.

As Langshans would become separated into two breeds, it is understandable that there was some variation in the apparent shapes seen at the shows up to 1900 or so. Most of the differences were in plumage quantity. Tighter feathered birds look taller, but may not really have longer legs than another, fluffier bird. Fanciers interested in any of the Langshan breeds as now bred should, if possible, examine as many illustrations, both drawings and photos, of Langshans before the divide.

CROAD LANGSHAN

The British Croad Langshan Club was formed in 1904, the new name or their type chosen to permanently honour the effective founder of the breed, Major Croad, and his niece. The Major had already died by then, and his niece died shortly afterwards, certainly by 1908. Croad Langshans soon built up fair support, if entries at the shows were any guide for those

Black Croad Langshans, pair. Artist: J.W. Ludlow. Originally a free gift with Poultry *magazine, circa 1912–14.*

days, when the shows were closely allied to commercial poultry keeping. Here are some Club Show entry numbers:

- 1910. Crystal Palace, 62 birds in 4 classes (cock, hen, cockerel, pullet)
- 1911. Crystal Palace, 101 birds in 4 classes
- 1912. Crystal Palace, 100 birds in 4 classes (10 c., 20 h., 38 ckl., 32 p.)
- 1913. Crystal Palace, 84 birds in 4 classes (8 c., 10 h., 39 ckl, 27 p.)
- 1921. Olympia, London, 107 birds
- 1922. Olympia, London, 146 birds
- 1923. Derby, 178 birds
- 1924. Olympia, London, 204 birds (probably the highest ever entry.) (22 c., 24 h., 45 ckl, 42 p., 33 novice, 38 selling)
- 1925. Birmingham, 155 birds.

This was probably the last of the really impressive displays of Croad Langshans in Britain, first hit by the General Strike, the Great Depression and then the Second World War. Croad Langshan Club Show entries were usually between fifty and seventy in the years from 1926 to 1939. They still managed to attract forty or so birds during the 1950s, falling to fifteen in 1966 and just six in 1968, when the Croad Langshan Club closed down. From 1969 until 1979, the breed was covered by the Rare Poultry Society, when interest perked up enough to reform the Croad Langshan Club.

The new club has certainly done a great deal to save Croads from extinction, but they have not attracted huge interest. Large black-plumaged chickens, no matter how useful they are, or how glossy the black, do not seem to be the first choice of newcomers to the fancy in the early twenty-first century. From 1980 to 2005, there have usually been about twenty large Croads at Club Shows, plus Croad Langshan Bantams (*see below*), that did not exist before.

Black Croad Langshan, large male. Photo: John Tarren

Black Croad Langshan, large female. Photo: John Tarren

Langshans

White Croad Langshan, large male. Photo: John Tarren

White Croad Langshan, large female. Photo: John Tarren

A few Blue and White Croad Langshans have appeared occasionally, for example, Miss A.H. Gordon Barrett, The Mount Farm, Farnham Royal, Buckinghamshire exhibited both colour varieties in 1930/31. Another fancier, Mr A. Powell of Essex exhibited a White Croad Langshan in 1937. They were not generally accepted by Croad Langshan breeders because they believed (probably correctly) that they had been produced by crossing with Blue or White Modern Langshans or Orpingtons. Miss Croad had made so much fuss about the 'purity' of Croad Langshans that it seems that her attitudes had lived on long after her death.

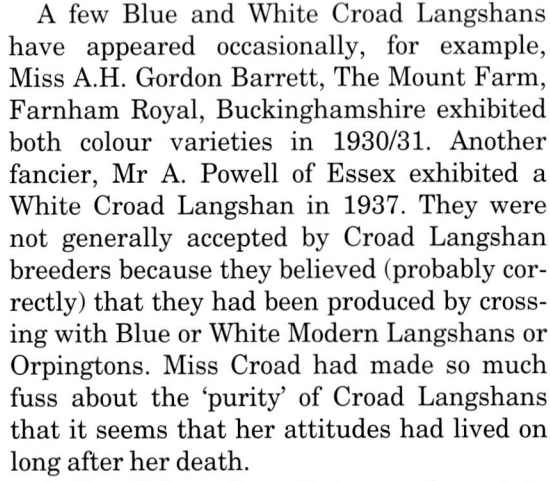

A Croad Langshan Club was formed in France in 1923. The author does not know how long it lasted. Possibly up to 1939? Croads lay brown eggs, some of a unique plum colour. They were used in the Netherlands to make Barnevelders. In Britain, Croads were used to make the first Black Orpingtons, and so the once commercially important Australorp. Thus Croad Langshans have probably been much more important as an ingredient of other breeds than they were in their own right. Commercially, they were never good enough as layers to tempt farmers away from Rhode Island Reds, and table bird producers wanted white or other light-plumage coloured
birds, not black ones.

Croad Langshan Bantam

The first reference the author has found to Croad Langshan Bantams was in the 1933 *Feathered World Yearbook*, where there was a mention in the text and an advert from the breeder, Mr C.N. Belbin of Nortonthorpe Hall, near Huddersfield, Yorkshire. It was generally assumed he had made them from crosses between Black Pekins and Black Old

English Game Bantams. As might be expected from this origin, they did not lay brown eggs, and were probably too small to really 'look the part'. This strain may have died out before the beginning of the Second World War in 1939.

According to Harold Easom Smith (*Modern Poultry Development*, 1976) a new, rather better, strain appeared at the 1954 National Show. This strain was much more successful, and still exists. It was partly made from under-sized large Croads, which would have helped the egg colour. A class of eight birds were exhibited at the 1962 'International' Show at the Olympia Exhibition Hall, London. This event was still a mixture of fancy poultry show and trade fair for the poultry industry. Wandelt and Wolters (*Handbuch der Zwerghuhnrassen*, 1998) mention a strain made from Black Booted and Australorp Bantams (and large Croads?). Since 2000 the Croad Langshan Club Shows have had entries in excess of twenty Black Croad Langshan Bantams some years, occasionally nearly thirty. There have also been some Whites, but no references of Blues have been found.

Description

The British Croad Langshan Standard requires the carriage to be: 'Graceful, well balanced, active and intelligent.' The first three are understandable, but it is not clear how judges are expected to assess their intelligence! Apparent 'Grace and activity' is achieved by leg length, and the amount and texture of body plumage, roughly halfway between stockier, fluffier Cochins and taller, tighter-feathered Modern Langshans. When standing normally, Croads have a roughly horizontal back, sweeping up to a large fanned tail. This gives the impression of 'bal-

Black Croad Langshans, bantam pair. Photo: John Tarren

ance', the tail appearing to be the same size as the neck and head when they are seen in profile. Croads have straight single combs, quite small for the size of the birds. Shanks and outer toes are lightly feathered. Large Croad minimum weights are 3.2kg (7lb) for females, 4.1kg (9lb) for males. Official Croad Bantam weights range from 650 (23oz) to 910g (32oz). In reality, they are much heavier than this, perhaps 900g (31oz) to 1400g (49oz).

Black Croads should have dark brown (not black) eyes, a horn coloured (not black) beak, shanks and toes that are nearly black on young birds, bluish-grey on older birds, with pinkish white soles and white toenails. Plumage surface colour must be black with a green sheen, but grey under-colour and a little white in the foot feathering is allowed.

White Croads should have as dark an eye and shank colour as possible. Blue Croads remain a non-standard variety, despite having been around since 1930. The Croad Langshan Club has probably not standardized them because of suspicions that they have been made by crossing with Blue Modern Langshans or Blue Orpingtons.

MODERN LANGSHAN

Mr R. Fletcher Housman was, along with Miss Croad, one of the people who reacted against the tighter feather and longer legs of some of the winning Langshans about 1890, and was one of the founder members of the Croad Langshan Club. His discontent was evident in a letter in the 12 February 1892 issue of *Feathered World* magazine:

> I noticed with regret while judging at the recent Liverpool Show that Langshan breeders seem to be getting rather astray from the standard of the breed in regard to the tails and legs of their birds. There is a noticeable difference in the size of the tails of a great majority of birds now, and the leg feathering is getting too sparse. . .'.

Mr Housman went on to (correctly) predict that the Langshan would follow the same path as 'Game' into separate Old English Game and Modern Game.

Miss Croad wrote her similar thoughts in a long letter in the 18 October 1901 issue of *Poultry* magazine. The key part is:

> At the time [1889–90] that the tall Game Langshan was introduced at the Crystal Palace Show, I met a disappointed exhibitor who pointed out to me a coarse, ungainly looking bird with which he told me he had won the challenge cup the previous year. He then drew me towards the very 'gamey' looking bird by which he felt himself to have been defrauded. On turning the matter over in my mind, I decided to write to the [poultry] papers. I stated that although I had no sympathy with the coarse-looking type of bird that had been winning the Langshan classes, I still considered it a very wide leap from the heavy bones and loose feathers of the Cochin to the tightness of the Game.

She also mentioned, generally favourably, the type of Langshan bred in America.

The Langshan Society, as the Croad Langshan Club after 1904, concentrated on Blacks. Blue and White Langshans, of the tighter-feathered 'Society type', were developed as well, but were listed separately in show results in the magazines.

Mr R.J. Pope of Barcombe, near Lewes, Sussex, developed one strain, starting in 1886 with two White pullets that he bred from some Black Langshans that had some white feathers. He had a flock of about seventy Whites by 1888 (ref. Lewis Wright, 1902). A Mr Will Smith produced another strain, by the same method, in 1896/7. Yet another strain appeared from faulty Blacks in the yards of Mr B.W.J. Wright in 1926.

Blue Langshans were made in the USA (*see* Langshan – US Standard), and sent over to Mrs Sismey of Rutland, who sold them on to Mr Shelton and Mr W.A. Jukes. Mr Jukes also bought some that had been sent over by

Black Modern Langshan male. Artist: M. Andrewes (circa 1900–14)

Mr Bradbury of New York, who had sent them over for the 1893 Liverpool Show – wealthy people did that sort of thing in those days. There was a separate Blue Langshan Club from 1906 to 1925, after which it merged with the Langshan Society. These were crossed with Society-type Blacks to produce Blue Langshans that were of the same tight-feathered type. Buff Langshans were mentioned in the 1880s, but had died out by 1900.

As Miss Croad had left the Langshan Society in 1890, it can be assumed that the Society catered exclusively for the tighter-feathered type. Society members did not accept the name 'Modern Langshan' until after 1945, despite the fact that everyone else had been calling them that since 1914.

'Everyone else' had decided that Croads must be a better utility breed than the Modern/Society-type, again despite the best efforts of the Society members to maintain their laying and table qualities. Therefore, the Modern Langshan was regarded as a high-class exhibition breed, and it is significant that when a specimen is seen at a show today, it is the poultry history enthusiasts who gather around in an excited manner to inspect them. The Langshan Society enjoyed a brief heyday, say from 1900 when all the Croad-type breeders had left, until 1914. They confined at a reduced level up to 1939, and existed in name only until 1969 when the Rare Poultry Society was formed. Their rise and fall should be documented in the hope that it will stimulate someone to establish another flock.

1901 'International Show' at Alexandra Palace, London, Blacks: 13 cocks, 23 hens, 53 cockerels, 65 pullets, 20 selling males, 25 selling hens.
Blues: 10 M., 8 F., Whites: 10 M., 16 F.

1910 Langshan Society Show at Sheffield, Blacks: 13 c., 15 h., 20 ckl., 29 p., Whites 6 M/F.

1910 Blue Langshan Club Show at Sheffield, 3 c., 11 h., 8 ckl., 9 p.

1911 Langshan Society Show at Leeds, Blacks: 7 c., 12 h., 30 ckl., 34 p., Whites 4 M/F.

1911 Blue Langshan Club Show at Leeds: 6 c., 6 h., 8 ckl., 10 p.

1912 Langshan Society Show at York, Blacks: 12 c., 22 h., 24 ckl., 26 p., Whites 5 M/F.

1912 Blue Langshan Club Show at York, 6 c., 10 h., 9 ckl., 12 p.

1913 Langshan Society Show at Crystal Palace, Blacks: 8 c., 12 h., 16 ckl., 18 p. No Whites.

1913 Blue Langshan Club Show at Crystal Palace: 3 c., 9 h., 13 ckl., 16 p. Harry Wallis, President of both the Society and the Blue Club, died in 1913. This, plus the fact that the same judge usually officiated in both sets of classes, indicates that friendly relations existed between them by this time.

1922 Langshan Society Show at Leicester, Blacks: 12 c., 20 h., 16 ckl., 16 p.

This seems to have been the last really good display of the Langshan Society, no doubt the economic depression playing a major part in their decline. The author has not found all of the Society Show entry figures between 1923 and 1939; those known range from twenty-eight to thirty-six. There are very few mentions of Blues or Whites.

Society members, especially Mr J. Stirzaker, repeatedly complained about non-specialist judges giving top awards to Langshans that were too tall and skinny, with tails that were too low and whipped. Some breeders had probably crossed them with Modern Game, which had taken the breed into an extreme form, far beyond the original differences between the membership which led to the separate Croad Langshan Club being formed. Mr Stirzaker acknowledged that young birds looked rather slender, but said that they were very heavy and meaty when mature.

The pre-war Society Secretary, John Pickerill, followed by his son Tom, kept the family flock going through until old age caught up with Tom, about 1970. This strain has survived through to the present (2005), mainly thanks to the dedication of successive generations of the Kay family, once of Newark, Nottingham, now at Alford, Lincs. John Kay was followed by his son Ian and now Ian's sons, Dean and Stuart. Modern Langshans have been covered by the Rare Poultry Society since 1969, and this breed has been kept by both Secretaries, Andrew Sheppy and Richard Billson. No references of Modern Langshans ever having been bred outside the UK have been found.

Modern Langshan Bantam

One would have expected that one of the many British fanciers who delighted in making miniature versions of almost every other large breed would have made Modern Langshan Bantams long ago, but no one did. Since the 1960s, a few fanciers have done so, including Ian Kay, Eric Parker (better known for his Polands) and Richard Billson. Breeds used to make them have included Black Booteds, Croad Langshan Bantams, Black and Birchen Modern Game Bantams. So far, they have failed to attract much interest outside the 'inner circle' of the Rare Poultry Society.

Description

The Langshan Society Standard, set in 1886, was never changed to properly describe the style of the birds that were winning, even at Langshan Society Shows, after about 1910. The Standard requires: 'Back horizontal when in normal attitude,' and 'Tail full, flowing, spread at base, carried fairly high but not squirrel, furnished with abundant side

Black Modern Langshan, large male. Photo: John Tarren

Black Modern Langshan, large female. Photo: John Tarren

hangers and two sickles, each tapering to a point.'

All the, no doubt carefully posed, champion Langshans photographed from 1910 up to the 1930s show birds standing well up, thus giving a sloping back and a lower carried, smaller, whipped tail.

The published standard does accurately call for 'close, compact plumage', 'legs rather long' and a light fringe of feathers down the shanks and outer toes. Although their tight plumage might make them look slim, they should be heavy birds, 2.7kg (6lb) pullets up to 4.5kg (10lb) adult cocks. Bantam males should be 1130g (40oz), females 910g (32oz).

Black Modern Langshans should have a very pronounced green sheen. The best specimens in their 'good old days' were said to be the equal of Cayuga and Black East Indian ducks in this respect. Blues should be laced as clearly as the best Andalusians, but the photographic evidence suggests they never were.

White Modern Langshans will probably need to be kept out of the sun to avoid brassiness. Most Whites seen throughout their history have been nearer the published standard in type than the 'fashionable' shape of the Blacks and Blues.

Eye, beak, shank, colours and so on, are all the same as for Croads.

LANGSHAN – US STANDARD

Langshans were first brought to America about 1877 or 1878, when Mr Samuels of Waltham, Massachusetts, imported some from Major Croad. Mrs Sargeant of Kettery, Maine imported many batches from the Major, effectively becoming his American agent. Pacific coast American breeders probably imported directly from China. Many of these details about Langshans in America are from articles by Joe Pehringer of West Allis, Wisconsin that appeared in *Poultry Press* magazine, 1988–90.

Black Langshans were admitted to the American Standard of Perfection in 1883, and Whites followed in 1893. American Langshan breeders were aware of the arguments and divisions going on among British Langshan breeders, and were determined not to follow the same path. The poultry shows were as highly regarded in the USA as they were in the UK, but seem to have had different priorities. Americans were already beginning to establish a modern poultry industry, and the shows were a shop window. British fanciers were still focused on show prize money and top prices for champions. Long-legged 'Society type' Langshans beginning to evolve in England would never succeed in the States. Croad type Langshans had several characteristics that were wrong for American table chicken producers, black plumage, white skin and feathered shanks; so they could never compete with White Rocks or White Wyandottes.

The American Langshan Club was formed at Boston, Massachusetts in January 1887. It had an initial boom up to 1900, with a number of wealthy members in New York State. These gentlemen gradually drifted off to other hobbies, to be replaced by more Midwestern members. This club covered all colours, but there was also an even more specialist National Black Langshan Club, formed in 1907, and having 170 members by 1913. The largest show entry details from this period the author has found were of the Baltimore Show, Maryland, 31/12/1912 to 4/1/1913, where there were sixteen cocks, twenty-five hens, twenty-one cockerals, twenty-one pullets and eight trios.

Some of the main early American breeders of Black Langshans included George Cooke of Racine, Wisconsin (who started in 1888), M.S. Barker of Thortown, Indiana (started 1902), Mr L.E. Meyer and his son W.A. Meyer, Bowling Green, Missouri (started 1896 and 1910), and Franklane L. Sewell, a name well known to poultry ephemera collectors as one of the artists who produced the now much sought after colour plates once given as free gifts with *Poultry Tribune* magazine.

White Langshans were imported from China by H.G. Keesling of San Jose, California and from England by Kirby & Smith, East Chatham, New York, both about 1889. Although never as popular as Blacks, White Langshans did enjoy a fair level of support up to about 1930. The largest show entry of Whites was believed to be at Pleasant Hill, Missouri in 1927, when eighty-four birds appeared.

Kirby & Smith were also involved in the development of Blues, first showing some at Charleston in January 1890. They had started from a blue 'sport' bred by a Mr McLean of Connecticut. Barreds, Buffs and Mottleds have all appeared at times, only to die out again.

The highest show entry of American Langshans the author has found was the 1986 American Langshan Club Show where there was a total of 213 entries. This included fifty-six large and thirty-one bantam Blues.

Langshan Bantam – US Standard

T.F. McGrew (*The Bantam Fowl*, 1899), told of a strain of Langshan Bantams made by accident. William M. Hughes had a large Black Langshan hen that hatched some chicks in November 1892 from a hidden nest. Five chicks survived the winter, which because of the cold weather never grew properly, only weighing about 1.5kg (3¼lb) at maturity. He bred them, selecting the smallest in successive generations. It is believed that he first exhibited them at the 1898 New York Show. This strain must have died out, as Black and White Langshan Bantams were not admitted to the APA Standards until 1960, with Blues only in the ABA Bantam Standard.

Description

American Langshans are generally similar to Croad Langshans, but more stylish. Much of this style, especially on males, comes from their large, fanned and highly carried tails. Body plumage is strictly 'medium', not loose like Brahmas or Cochins, or close like Modern Langshans. Weights range from 3kg (6½lb) pullets up to 4.3kg (9½lb) adult cocks for the large fowl, with the miniatures about 800 (28oz) to 1,000g (35oz). Most Langshan Bantams in America will probably exceed these weights, despite the usual American strictness on standard weights.

CHINESE LANGSHAN – AUSTRALIAN STANDARD

Australian livestock breeders have always been somewhat isolated, in the nineteenth century by the long travelling distances involved, more recently by laws that forbid most imports, and place strict quarantine regulations on any that are allowed.

Despite all this, there were importations of Langshans in the late nineteenth and early twentieth centuries, probably some direct from China, Croads and Moderns from England, and possibly even American and German types as well. In addition to all these, there were the first, utility-type Black Orpingtons (originally partly made from Langshans by William Cook), that would eventually return to England as the 'new' Australorp. Work at Hawksbury Agricultural College played an important part in creating the standard 'Chinese Langshan' that might more accurately be called 'Australian Langshan' from this mixture. A breed club was formed in 1920. There are Blues and Whites as well as Blacks.

Chinese/Australian Longshan Bantam

These were first made during the 1950s by Gordon Everleigh of Gunnedah. He initially crossed Black Pekin and Black Modern Game Bantams, then added large Australian Langshans. Since then, Blue and White miniatures have been added to the range.

Description

Large Chinese/Australian Langshans are the smallest of the types of Langshan, only expected to weigh 2.5kg (5½lb) (hens) to 3kg (6½lb) (cocks). The back and tail formation is very similar to that of Deutsches Langschans and British type Australorps. Neck hackles sweep well down the back, to a point where the saddle starts to rise up to the tail, thus appearing to have almost no 'back' at all. They have shorter legs than any of the other types. Colour of eyes, shanks, and so on, are as for the others. Australian Langshans are said to be among the best layers and table birds of the family, and with their more modest size, are perhaps the most practical of the group for those who are interested in self-sufficiency as much as showing.

DEUTSCHES LANGSCHAN (GERMAN LANGSHAN)

The first Langshans in Germany and Austria-Hungary came from Major and Miss Croad in England in 1879. Croad Langshans are still bred today in Germany, and probably in the now separate Austria and Hungary alongside the German Langshan breed that was shortly to develop as well. Some breeders were said to have crossed them with Plymouth Rocks and other breeds that would have started the process towards the Deutsches Langschan.

Baron Villa-Secca of Vienna imported some White Langshans in 1880, and then crossed them with White Orpingtons that were then very new, and nothing like today's 'Orpington Shape'. They had tighter body plumage and large tails, still showing a lot of the White Dorking that William Cook had used to make them. The Baron went on to develop a strain of birds to his liking that would become the beginning of the tall, clean-legged birds with a rising saddle that has become the Deutsches Langschan. It may be the case that the black (and blue?) Langshan crosses mentioned above could only be given a proper name, 'Deutsches Langschan', after an influential person, such as the Baron.

The 'Clubs Deutscher Langschan-Zuchter' was formed on 12 September 1895. As with

Black German Langshans, large pair. Artist: Kurt Zander

the other types of Langshan, Blacks became the most popular colour, followed by Whites, then Blues. Barreds have been made very recently (circa 2000–05). Other colours, Birchens, Buffs and Brown-Reds are only seen as bantams.

Deutsches Zwerg-Langschan/ German Langshan Bantam

Rektor Johann Heermann, of Wedel, Holstein, started to make Croad-type Langshan Bantams in 1902 from large Langshans, Black Cochin Bantams, Black Rosecomb Bantams, and Silver Duckwing-coloured 'country bantams'. He first exhibited some in the summer of 1910 at Hamburg, By this time, the large, tall, clean-legged Deutsches Langschan was being established, so in 1911 he added Birchen Modern Game Bantams to the mix.

About 1916, Reds appeared, made from Black Deutsche Zwerg-Langschans, Buff Cochin Bantams, Wheaten OEG Bantams and Rhode Island Red Bantams, the last-named probably being the most important, as Rhode colour was the intended ideal.

Josef Schäfer of Aschaffenburg entered the first Laced Blues (Andalusian colour) in 1924 at the 5th Nationalen Zwerghuhnschau. Whites were also made during the 1920s. As with all the German breeds, progress was halted and much was lost during the disruption of, first the hyper-inflation years, then the rise and eventual fall of the Nazis.

Starting about 1956, and producing showable birds by 1962, Dr Günter Paetz of Wolfenbüttel made Brown-Reds and Martin Hoffmann of Gladbeck made Birchens. Later in the 1960s, Hans Konze of Rietburg made Barreds, using

Langshans

Barred German Langshans, large pair. Artist: Kurt Zander. Clearly a strain of these was attempted in the 1920s, for Kurt Zander to illustrate them, but it was believed that they did not survive long. Current stock are a recent creation.

Blacks and Barred Plymouth Rock Bantams. During the 1990s another batch of colours have been added to the range, first Lemon-blues (blue version of Brown-Reds) and Silver-blues (blue version of Birchens) then, later, Buffs, Columbians and Buff-Columbians.

Description
German Langshans have often been described as having a 'wine glass' outline when viewed in profile, their long legs being the stem, and the curve of the well-rounded breast being as symmetrical as possible with the belly. The top of the head should be the same height as the tip of the tail when correctly posed, thus helping the wine glass analogy. There is a convex curve on the neck hackle that is mirrored on the curve of the rising saddle. When viewed from above, there is a definite 'correct' width to the rising saddle, neither too narrow or too wide. Large German Langshans weigh from 2.5kg (5½lb) up to 4.5kg (10lb) adult cocks. Bantam females weigh 1,000g (35oz), males 1200g (42oz), an increase on previous standards of 900g (31oz) and 1,000g (35oz).

Their single combs are small, single, straight and upright. Everything about German Langshans is neat and tidy. Eyes are dark brown on Blacks, Blue, Brown-Reds, Birchens, Lemon-Blues and Silver-Blues; orange-red on the lighter colours. Shanks and feet are black on young birds of the dark varieties, slate on older birds. Barreds have white shanks/feet, and the other lighter colours have light slate shanks

Langshans

White, Blue, Black and Red German Langshans bantams. Artist: Kurt Zander

Blue German Langshan, large female. Photo: John Tarren

Black German Langshan, bantam male. Photo: John Tarren

Blue German Langshan, bantam female. Photo: John Tarren

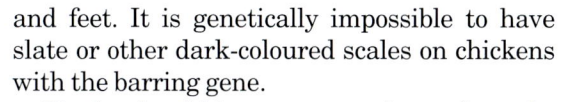

White German Langshan, bantam male. Photo: John Tarren

and feet. It is genetically impossible to have slate or other dark-coloured scales on chickens with the barring gene.

Blacks should have a green sheen throughout. Blues should be laced, as Andalusians. Whites should be snow-white, and will need appropriate care to keep them so. Brown-reds, Birchens, Lemon-Blues and Silver-Blues should all be as Modern Game. It is assumed by the author, who has not even seen pictures of Buffs, Columbians or Buff-Columbians, that they are similar to other breeds with these varieties.

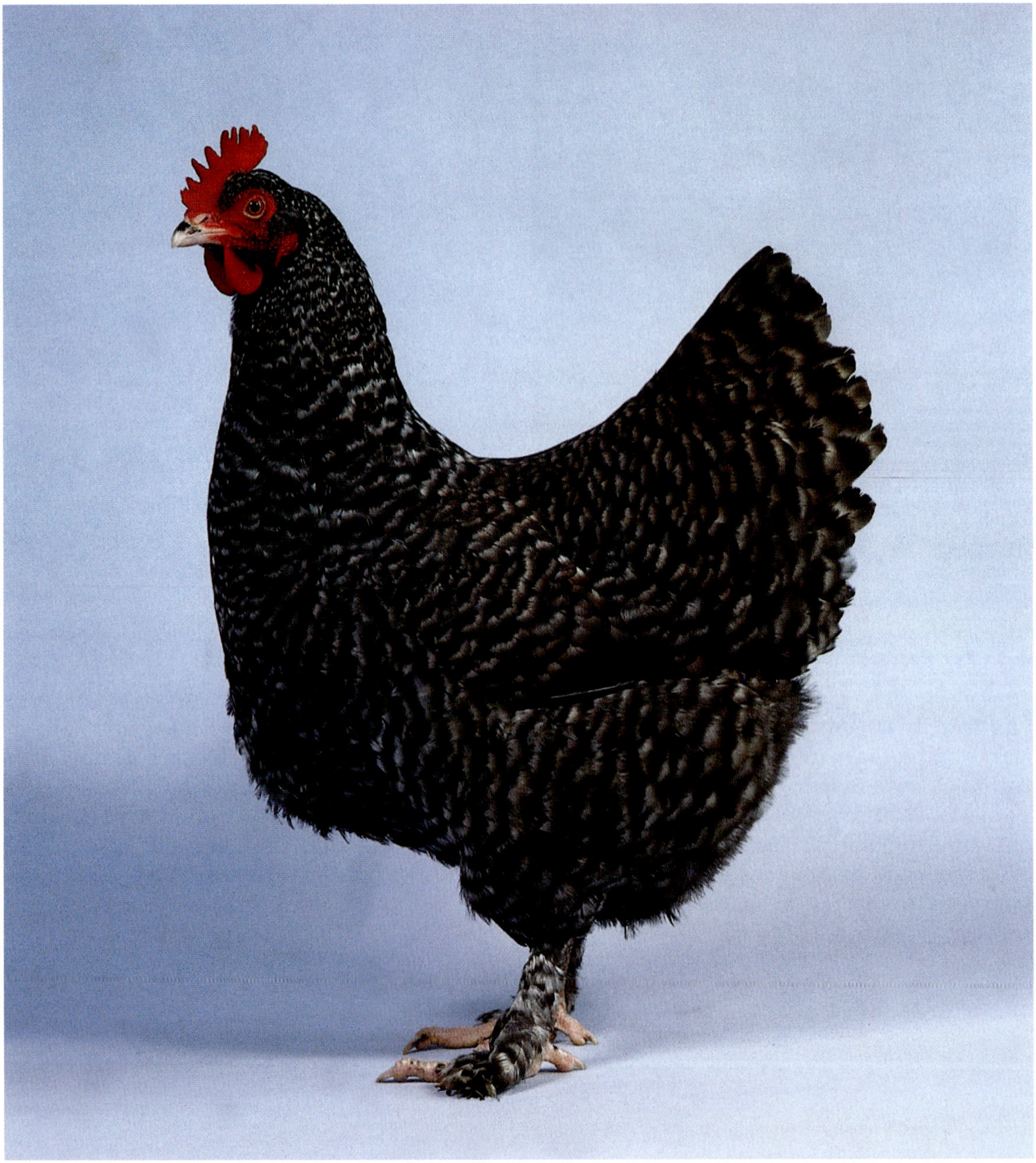

CHAPTER 11

European Medium/Heavy Breeds

Many of these were developed from crossing more ancient European types with the heavy Asiatic Cochin/Shanghai fowls imported from about 1850 onwards. It naturally took some time for reasonably uniform new breeds to be established from initial crossbreds. Some breed clubs were formed between 1880 and the outbreak of the First World War. The definitions of 'light', 'medium' and 'heavy' breeds are sometimes arbitrary.

Breeds of this group that are not included in this book:

(1) Those too popular to be considered 'Rare': Australorp (nominally Australian, but having a joint origin with Orpingtons), Barnevelder, Dorking; Faverolles (including Deutsches Lachshuhn and Faverolles Claire); Marans (French and British types), Orpington, Sussex; Welsummer/Welsumer (classed as a 'light breed' by PCGB, medium elsewhere).

(2) Those too new and/or too closely related to other, more popular breeds:
Amrock (Utility-type Barred Plymouth Rocks made into a separate breed in Germany 1958); Aquitaine (French variant of Black Australorps circa 1968); Charollaise (in 1966 French Standards, from crosses of Schweizerhuhn and Gâtinaise); Dorset (British utility breed, never recognized by PCGB or RPS).

(3) See Autosexing Breeds chapter for: Bielefelder Kennhuhn, Brockbar, Buffbar, Dorbar, Polbar, Rhodebar, Welbar, Wybar.

Breeds covered in this chapter:

Barbezieux – France
Bourbonnaise – France
Bourbourg – France
Coucou de Flandres/Vlaanderse Koekoek – Belgium
Coucou d'Iseghem/Izegemse Koekoek – Belgium
Coucou de Rennes/Coucou de France/Coucou de Bretagne – France
Deutsches Reichshuhn – Germany
Dresdner – Germany
Estaires (or D'Estaires) – France
Gâtinaise – France
Ixworth – England
Lincolnshire Buff – England
Malines/Mechelse Hoen – Belgium
Mantes – France
Neiderrheiner – Germany
Noire du Berry – France
Noire de Challans – France
Noord-Hollandse Hoen – Netherlands (clean-legged)
North Holland Blue – England (feather-legged)
Norfolk Grey – England
Schweizerhuhn – Switzerland
Sundheimer – Germany

European Medium / Heavy Breeds

BARBEZIEUX

This can be regarded as the French version of the Minorca, as it is a heavy-weight, black plumaged bird of Mediterranean type with white ear lobes. With weights of 3.5kg (7½lb) for females and 4.5kg (10lb) for males (and these have been exceeded), they can be very large indeed. Their place of origin is the town of Barbeziex in the Département Charente in western France, south-east of Cognac. As far as their history is concerned, they were briefly mentioned in Lewis Wright's *New Book of Poultry* (1902), but not in the earlier editions, and J.T. Brown's *Encyclopaedia of Poultry* (1908), and in Edward Brown's *Races of Domestic Poultry* (1906) they were in the chapter on Spanish breeds because Mr Brown (who later became Sir Edward) thought their relationship to other breeds from Spain was more significant than the fact that they were bred in France. The difference in approach to poultry breeding in France and Britain in the nineteenth century, and its consequences for twenty-first century hobbyists, is clearly seen if the history of Barbezieux is compared with that of Minorcas. Barbezieux were a more practical breed, but never spread beyond their home region. Minorcas, developed at least as much for showing as eggs or meat, are bred in many countries around the world.

BOURBONNAISE

From the central French Département Allier, probably around 1900, they were not mentioned by Edward Brown in *Races of Domestic Poultry* (1906), but were in J.T. Brown's *Encylopaedia of Poultry* (1908). They still exist (2005), and look very similar to utility-type Light Sussex, smaller and more active than the exhibition strains. Standard weights range from 2.5kg (5½lb) females up to 3.5kg (7½lb) males. They were probably the result of imported Light Sussex and/or Light Brahmas

Black Barbezieux, quartet. From a series of trade cards issued by a number of chocolate and coffee shops in France. Artist amd date unknown, possibly 1890s.

being crossed with local farmyard hens, possibly birds similar to Bresse. Bourbonnaise have grey undercolour, a distinction from the white undercolour of Light Sussex.

BOURBOURG

These are also a Light Sussex/Light Brahma colour French breed, but are larger and have lightly feathered feet. Apart from not having a beard or five-toed feet, they are very similar to Ermine Faverolles. They originate from the town of Bourbourg in the Département Nord, between Calais and Dunkerque and were bred to supply eggs and table birds for much of north-western France by 1900.

Bourbourg Fowls remained a popular practical breed until replaced by hybrid broilers and layers in the 1950s. Fanciers and conservationists began to revive them around 1975. The author does not know if current strains are completely new, or if there were at least a few survivors from pre-war stocks involved. It is to be hoped that the Bourbourg enthusiasts in France maintain their strain as this breed is unlikely to replace Light Sussex or Ermine Faverolles among fanciers in the UK and elsewhere, or the also very similar Sundheimer in Germany.

COUCOU DE FLANDRES/ VLAANDERSE KOEKOEK/ FLEMISH CUCKOO

Virtually nothing has been recorded of the early history of this breed, or its close relations described below. They were just the local chickens of Flanders, along with their neighbours in France. There is probably also a relationship with Scots Greys, although these were eventually developed into a slimmer, sharply barred breed, unlike the cuckoo barred breeds of Belgium, France and The Netherlands, all of which became heavier types.

Coucou de Flandres were one of the direct ancestors of Malines, the original type before they were made larger by crossing with Antwerp Brahmas and other Asiatics from the 1850s onwards. Standard weights from 2.5kg (5½lb) pullets up to 3.5kg (7½lb) adult cocks have been required since about 1900, but actual weights were probably lower before 1850.

The breeds was promoted as something distinct and separate from Malines by Messieurs A. Detroy and Edouard Labbe in Belgium during the 1890s, but the breed struggled to maintain its separate identity beyond the First World War. Local enthusiasts revived Coucou de Flandres again in the 1980s, but the author does not know how many, if any, of the original strains had survived for them to work from. Wandelt and Wolters (*Handbuch der Hühnerrassen*, 1996) show photographs of present-day Coucou de Flandres. To fully fit the descriptions of the originals, they need larger (single) combs and larger tails, carried higher. No references to a bantam version have been found.

COUCOU D'ISEGHEM/ IZEGEMSE KOEKOEK

These are named after a small town in south-western Belgium, a short distance north of the city of Courtrai/Kortrijk. They were probably really bred over a much wider area, as their name is simply used for the rose-combed version of Coucou de Flandres. Coucou d'Iseghems enthusiasts claim that the breed can be traced back to the year 1554.

The effective recorded history of the breed started with an article in 'Chasse et Peche' magazine in 1907. This encouraged several local breeders to form a breed club, the 'Hoenderbond T'Neerhof van Izegem'. They published the first standard for the breed on 12 February 1909. Club Shows were held from 1910 until 1914, and then again after the First World War, restarting in 1921. Some were displayed at the Second World Poultry Congress, Barcelona, Spain, 10–18 May 1924, but they failed to attract much international interest. Coucou d'Iseghems continued to be bred in considerable numbers in their home region until depleted by the disruption of the Second World War.

Current strains are believed to be a remake, developed by a group of historically-minded fanciers from 1966 onwards. Cuckoo Malines and Bruges type Belgian Game were two of the breeds used. It is not known if any of the original strain survived at all.

The present breed standard, accepted in 1978, gives weights from 3kg (6½lb) pullets up to 4kg (8¾lb) for adult cocks, both 500g (17oz) lower than the 1909 standard. No doubt present day breeders would be more than pleased if they reared any up to the old weights. These large birds are active, with a fair length of leg. Their tails are quite short and held about 30–45 degrees above horizontal. They have clean (not feathered), white shanks and feet. Combs are rose, fairly low, with a leader (rear spike) which ideally follows the line of the skull. Specialist poultry artist Rene Philippe Henri Delin (2.8.1877–15.9.1961) produced two illustrations of the breed that are regarded as templates for breeders: a drawing in 1923 and a watercolour in 1944, the latter usually found as one of a series of seventy-two postcards published by the Royal Belgian Institute for Natural Sciences. No records of a bantam version of this breed have been found.

COUCOU DE RENNES/COUCOU DE FRANCE/COUCOU DE BRETAGNE

These are from north-western France, near the border with Belgium. The name Coucou de Rennes is traditionally given to single-combed birds, Coucou de France or Coucou de Bretagne for those with rose combs. All have larger combs and slightly lower body weights than Belgian equivalents, Coucou de Flandres and Coucou d'Iseghem. As said above, they all have a common ancient history. Edward Brown named and defined these birds in these terms in *Races of Domestic Poultry* (1905), but it is not known how long this had been the case. Before 1850 they would have simply been anonymous chickens.

DEUTSCHES REICHSHUHN

As the name suggests, there was an element of nationalism in the motives of the breeders in the creation of the Deutsches Reichshuhn breed. The first colour variety, White, was made by 1907 or 1908, with their development believed to have started about 1895. They were made by crossing rosecombed White Dorkings, Columbian Wyandottes, White Wyandottes, White Orpingtons (rosecombed strain from USA), White Minorcas, Light Sussex, Cuckoo Malines, and Dominiques. Early illustrations suggest that these first Deutsches Reichshuhner were deeper-bodied than the elegant birds they are today. The originators of the breed were Hauptmann (Captain) Karl Cremat, his daughter, Frau Elli Schreiner and Willheim Onken.

Deutsches Zwerg-Reichshuhn/Deutsches Reichshuhn Bantam

Heinrich Borgards of Mülheim/Ruhr and Richard Knieper of Wilhelmshaven started to make the miniature version of the breed about 1930, showing their prototypes at the 1932 Hamburg-Wilhelmsburg Geflügelschau. They started with the 'Hell' variety (= Light, as Sussex), using undersized large fowl of the same colour and Columbian Wyandotte Bantams. Albert Winkler of Mühlau displayed the first White miniature in 1935 at Leipzig. The other colours had to wait until the 1950s and 1960s. These developments continued in both the DDR (east) and FDR (west), officially separately. Despite being very elegant little birds, arguably prettier than the full-sized version, they are seldom, if ever, seen outside Germany. At the main German shows, entries of well over 100 Deutsches Reichshuhn Bantams can be seen.

Description

They are at the low end of the medium/heavy breeds as far as size is concerned. Large pullets are a modest 2kg (4½lb), up to 3.5kg (7½lb) adult cocks. Bantams range from 800g

White Deutsches Reichshuhn, large male. Artist: August Rixen

White Deutsches Reichshuhn, large female. Artist: August Rixen

(28oz) up to 1,000g (35oz). Deutsches Reichshuhner have a long flat back, accentuated by their tail carriage, only slightly above the horizontal. Their rose combs are compact, with the leader following the skull as on Wyandottes. Unlike Wyandottes, they have white skin, shanks and feet.

The standard colour varieties are: Barred, Birchen (solid black breast, as Norfolk Grey), Black, Buff Columbian, 'Hell' (Light, as Sussex), Red (as Red Sussex, lighter than RIR) and White.

DRESDNER

Dresdners are very similar to Deutsches Reichshuhner in several respects – they are also from Germany, and are also partly made from Wyandottes. They are relatively new by the standards of pure breeds, the first stage of their development being from 1948 to 1953. Alfred Zumpe produced them from crosses of Rhode Island Reds, New Hampshire Reds and utility-type Wyandottes. The first colour variety was light red, very similar to New Hampshires. Some readers might need to be reminded that Dresden was in Communist East Germany, the DDR. Three more colour varieties were made between then and 1981: Black, White and Black-Red/Partridge. They are popular in Germany, especially the 'Goldbraun' variety, but seldom seen elsewhere.

Zwerg-Dresdner/Dresdner Bantam

The miniature version of the breed was developed by Siegmar Zumpe, the son of Alfred, the originator of the large fowl. They lived about 20km (12 miles) east of the city of Dresden. The process started in 1951, before large Dresdners had been perfected, so the two sizes were almost made simultaneously. Despite the division of Germany (the 'Iron Curtain', Berlin Wall, and so on) they soon became very popular in West Germany as well. As with several other breeds in Germany during this period, the evidence suggests that poultry fanciers had much more contact between west and east than anyone else (friends in high places?). K. Heilmann from Osnabrück produced Blacks in 1969, followed by Whites in 1976, made by J. Kraus (Norsingen) and E. Edel (Rheinfelden). As with the large fowl, Dresdner bantams are popular in Germany but virtually unknown anywhere else.

Description

In general terms, Dresdners could be described as slimmed-down and stretched-out Wyandottes. Comb shape and yellow skin, shanks and feet are all as Wyandottes. 'Goldbraun' is similar to the colour of a New Hampshire Red. The Black-Red/Partridge is black breasted, but otherwise similar to Welsummer colour, darker than many other breeds with this plumage variety. Whites and Blacks are as normal. Because they have yellow shanks/feet, they may need the same double mating (Blacks) and feeding regime (Whites) necessary with the same colour varieties of Leghorns and Wyandottes.

ESTAIRES OR D'ESTAIRES

Estaires is a village in the Département du Nord, the area of north-west France, near the border with Belgium. Table poultry producers in this region traditionally crossed their normal farmyard hens with the giant fighting cocks of the area, 'Grand Combattant du Nord'. According to T.R. Robinson (1929 *Feathered World Yearbook*), Langshans were bought from Major Croad and bred by M. de Foucard and M. Robert Fontaine, and from them they became available to others to add to the mix. According to Edward Brown (*Races of Domestic Poultry*, 1905) the Estaires breed became a regular sight in the markets of Lille, the main city of the region, by about 1890. The breed was mainly bred to the west of the city, Merville and Lavente being two other small towns associated with the breed. Unfortunately for the people, livestock, buildings and everything else there, this was the area devastated by the trench warfare of the First World War. Enough survived to see them through to 1939, but the author thinks they had died out by 1945. A new strain was made by fanciers about 1980 using Grand Combattant du Nord and Croad Langshans as before, with (French type) Marans taking the place of the unspecified local hens involved a century before.

Estaires look like what they essentially are: Croad Langshan crosses. They are large, black-plumaged, white-skinned and with lightly feathered shanks; differing from Croads by having red eyes and slate-blue shanks/feet. The author does not know about the current stocks, but in their heyday (1890–1914) many of the cockerels had gold or silver neck and saddle hackle feathers. As they were destined for the table, not the show hall, no one cared. As far as is known, there has never been a miniature version.

GÂTINAISE

These are a white-plumaged table breed developed to supply the Paris markets from the areas to the south and east of the capital. A. Renard was the first recorded exhibitor of the breed at the 1906 Paris Show, but they may have existed as anonymous white chickens for a long time before then. A Gâtinaise Club was formed in 1909 that is believed to have been active at least up to 1939. The largest show entry recorded was 231 specimens at the 1934 Paris Show. Wandelt and Wolters (*Handbuch der Hühnerrassen*, 1996) have a modern photograph of a cockerel, evidence that they still exist. As far as the author is aware, they are not bred outside France.

They are large birds, ranging from 2.5kg (5½lb) pullets up to 4kg (9lb) adult cocks, some of which exceed 4.5kg (10lb). The comb is single and upright, lobes are red, plumage, skin, shanks and feet are all white. British fanciers are unlikely to take them up as they are almost identical to White Sussex, already established here.

IXWORTH

The originator of this breed, Reginald Appleyard, from the village of Ixworth, Suffolk, England, is perhaps better remembered for his other creation, the Silver Appleyard Duck. His chicken breed, the Ixworth, was intended to be a specialist meat breed. Starting in 1931, he intended to make a white-skinned, white-plumaged, heavily fleshed breed that would avoid the problems associated with Indian Game. These were infertility, yellow skin and breast blisters caused by the scanty plumage of Indians. Indian Game hens are very poor layers, so many British table chicken producers relied on Indian (Dark or Jubilee) Game cocks mated to Light Sussex or Buff Orpington hens to produce commercial meat chicks.

Jubilee Indian Game and White Sussex were the two main varieties used, but Dark Indian Game, White Old English Game, White Orpingtons and White Minorcas were also involved. He first exhibited his new breed at the 1938 London Dairy Show, then still as much, if not more, a trade fair for farmers and butchers as a fanciers' event. In addition to classes for live poultry, sheep, pigs and cattle, there was a large section for carcasses – in the case of poultry often in trade boxes of six or more birds. The Poultry Club of Great Britain accepted Ixworths as a standard breed in 1939, another reminder that poultry showing was still connected to commercial poultry farming then.

The industrial-scale poultry industry we have today was beginning to develop at this time in the USA and Canada. It was estimated (Dr E.W. Benjamin, 1939 World Poultry Congress) that 25 per cent of the world's egg and poultry production was in the USA. Scientists were applying their knowledge to farming in general, chickens in particular, to improve housing, nutrition, disease control and were making progress from a zero start in the brand new science of genetics. Although rail transport had been available for several decades, this was supplemented in the 1930s by improved refrigeration equipment and bigger, faster, more reliable trucks for road transport.

Poultry farmers in Britain and mainland Europe were not able to develop their production methods along the same lines because of the disruption of the Second World War. Reginald Appleyard was not the only British breeder to foresee the broiler chicken; Mr and Mrs Henson made 'Peterboroughs', Mrs Jenkins of Longparish, Hampshire made 'White Surreys' along similar lines at the same time. By the time peace was restored in 1945, large-scale American breeding companies had made so much progress in improvements in growth rates and feed conversion ratios that these three small-scale British breeders were never going to be able to catch up. American broiler strains therefore dominated table chicken production all over the world from the 1950s. Peterboroughs and White Surreys both died out completely, but against all expectations a few Ixworths survived through to the 1970s, when they were revived by members of the historically minded Rare Poultry Society.

Since the 1990s, a number of specialist producers in the free-range/organic sector have taken up Ixworths, so their current numbers are probably considerably higher than the very few seen at poultry shows would suggest. Being simply large white chickens, Ixworths look too much like broilers to be of interest to many show exhibitors or those who simply like to keep a flock of ornamental layers in their garden.

Ixworth Bantam

Reginald Appleyard also made Ixworth Bantams, starting about 1945. He crossed several breeds of bantams: White Rosecombs, White Old English Game, White Sussex and Jubilee Indian Game. It is believed he first exhibited them in 1949, probably at the London Dairy Show. Very few bantam fanciers were interested, so they died out; possibly surviving until about 1965.

Ixworth, large male. Photo: John Tarren

Ixworth, large female. Photo: John Tarren

Description

The tendency is for Ixworths to look too much like a white-skinned White Indian Game. Certainly they should have pea combs and some Indian Game character, but Reginald Appleyard deliberately avoided going too far in this direction. Their legs should be long enough for them to be active and fully fertile, and their shanks should be strong, but not as thick as those of champion Indians. Plumage is short and close enough to give a good impression of their body shape, but not as sparse as that of Indians. Rapid and adequate feather growth helps to reduce food consumption and minimize breast blisters and dented breast bones on growing table cockerels. British Poultry Standards require pullets to weigh 2.7kg (6lb), adult cocks 4.1kg (9lb) but many exceed these weights. The bantams were supposed to weigh 790 (28oz) to 1020g (36oz) according to age and sex, but were probably really larger than this.

LINCOLNSHIRE BUFF

The Lincolnshire Buffs seen at some British shows today are from a 'facsimile' strain that was developed in Lincolnshire during the 1980s. The originals were recognized by the meat trade in the late nineteenth century, but were never standardized by the Poultry Club of Great Britain.

Before the 1850s, poultry keeping in Lincolnshire was a small-scale affair, just intended to supply the local population. Table birds were bred by mating assorted local hens with Dorking and/or Game cocks. After 1850, another heavy breed became available, the Buff Shanghai, later standardized as the Buff Cochin. The first imports were not as profusely feathered as Buff Cochins have been known throughout the twentieth century, but were similar (in plumage quantity) as Croad Langshans. Similar Buff Shanghai/Cochin crosses were used to create several other breeds around the world, including several of the American breeds. In all cases, including in Lincolnshire, it took many years before a uniform 'new breed' was created from the crossbreds.

There had to be a reason for the development of the Lincolnshire Buff in large enough numbers to become more than just the occupants of a few isolated farmyards. In this case, it was the growth of the railway network. Before their railway stations opened,

Boston, Sleaford and Spalding markets were just that; but with the trains, the London markets suddenly became readily accessible.

They did not care about the exact shade of plumage. 'Buff' in the broadest sense of the word was good because they plucked out 'clean', without the dark feather stubs that would be a problem on Black or dark Black-Red/Partridge birds. Some light partridge markings would not have been a problem, nor were black tails. Some had four toes, others five (from the Dorking ancestors first used in the area), this did not matter then. Some had feathered shanks from the Shanghais, others were clean-legged, from the earlier types (Game, Dorkings and their crosses do not have feathered shanks). In this case, clean shanks were favoured as practical poultry people already knew that feather-legged chickens are prone to scaley leg mite. Some had yellow skin, shanks and feet (from most of the Shanghais), others had white skin, and so on. Most British consumers preferred white-skinned table birds, yellow 'corn fed' chickens were then only favoured in America. Mr R. Seed of Tattershall, Lincolnshire, when describing his local breed to the great poultry expert and author, Harrison Weir in 1895, said they were by then as uniform as they needed to be for table bird production.

William Cook launched his Buff Orpingtons in 1894. He was a great publicist as well as a very large-scale breeder by the standards of the time. Buffs were the third colour of Orpingtons he made, following the already popular Blacks and slightly less popular Whites. For the first decade or so, Buff Orpingtons were much tighter in feather than the giant fluffy birds they are today; very similar to Lincolnshire Buffs in fact. From the mixed bag that Lincolnshire Buffs were, a fair proportion could be found that 'would pass' as Buff Orpingtons. A Lincolnshire farmer was quoted in *Feathered World* magazine (16/12/1898): 'If I call my birds Lincolnshire Buffs I can't get more than four shillings each for them, but if I call them Buff Orpingtons they sell readily at ten shillings each.' It was alleged that many poultry dealers regularly visited Lincolnshire farms to choose the birds that suited them. William Cook always denied doing so, but, as Mandy Rice-Davies so famously said, 'He would, wouldn't he.'

Black Orpingtons were already well-established as show birds, and were fetching remarkable prices by 1894. It was a simple matter to establish Buffs as show variety, although some were quick to point out that the Cooks had made Blacks and Buffs from completely different mixes of breeds. When researching for this chapter, the author only found one class for 'Lincolnshire Buffs' at a show, that being the Lincolnshire County Show at Brigg on 26 July 1901. This was a class for 1901 hatched pairs. There were four pairs entered, first prize going to Lady Winefride Cary-Elmes. It was noted that this pair were rosecombed. Second prize went to Mrs Tumill and third to Mrs Tindale – the owner of the fourth pair was not named. This was reported in *Poultry* magazine that also carried adverts for several Lincolnshire Buff breeders during 1901. Most were in Lincolnshire, but there were also adverts from Stratford, Essex, and Frome and Yeovil in Somerset. It is difficult to be sure when the original Lincolnshire Buffs died out, probably a few years after the end of the First World War.

The project to make new Lincolnshire Buffs started in 1981 at the small animals unit of the Lincolnshire College of Agriculture and Horticulture at Riseholme. The college's involment in the process ceased in 1986, but by this time a number of hobbyist breeders had joined in. Brian Sands of Friskney, near Boston, was the prime mover. By 1994, there were enough people for them to start thinking about forming a 'Lincolnshire Buff Society'. It was formed on 29 January 1995, but neither the Society nor the breed could be recognized by the PCGB until the breed had been standardized. As the breed had not been standardized before, the enthusiasts drew up a standard based on the existing records, most of which were from Harrison Weir. This

European Medium/Heavy Breeds

process was carried out by the Rare Poultry Society, of which Brian and most of the others involved were members already. The standard was passed and the Society officially recognized in January 1997. Since then, the Society has attracted about sixty members, the majority of which live in or near Lincolnshire. There have never been any Lincolnshire Buff Bantams; and as there are many buff or similar coloured bantams already, Nankins for example, there would be little point in making them.

Description

Despite having sixty society members, very few Lincolnshire Buffs are seen at poultry shows. Most of the breeders probably keep them because if they are going to keep chickens, they may as well have the local breed. The poultry industry has long since severed all contact with poultry showing, but hobbyist poultry keeping is very much influenced by the Shows. Hence the need for a standard.

In keeping with their original function, Lincolnshire Buffs have to be large and meaty. Adult cocks should weigh at least 4.5kg (10lb), hens 3.6kg (8lb); cockerels and pullets about 1kg (2lb). They do not look as large as Orpingtons because they have tighter plumage that combined with their longer tails, makes them look quite similar to Sussex in general shape. Plumage is several shades of buff, lightest on the belly and breast, darkest across the shoulders. Their tails are bronze or brown, not black. Lincolnshire Buffs have now been standardized with five toes, which, as said above, some of them had before. Skin, shanks and feet are white, which was preferred if not always obtained before.

MALINES/MECHELSE HOEN

The French name, Malines, is generally used internationally rather than the Flemish name Mechelse Hoen, for this Belgian breed. Malines/Mechelen is a town to the north of Brussels. The breed was developed, like many others around Europe and America, from imported Asiatic fowls being crossed with pre-existing local breeds. A Mr de Winter was credited as the founder of the breed, his initial breeding programme running from 1850 to 1860. We cannot expect to know exactly what matings Mr de Winter employed

Lincolnshire Buff, male. Photo: John Tarren

Lincolnshire Buff, female. Photo: John Tarren

some 150 years ago. We are lucky to know any details at all. Mr de Winter obtained his Asiatic fowls from Antwerp Zoo, initially in 1850. Cuckoo Cochins are mentioned in some sources, as are 'Antwerp Brahmas'. Did the Zoo have both?

'Antwerp Brahmas' (or 'Brahma d'Anvers') were described by Edward Brown (*Races of Domestic Poultry*, 1905). They were large birds, considered a valuable meat breed. To suit European preferences, they had white flesh, skin, shanks and feet. They had feathered legs and feet, probably quite light, like Croad Langshans today; in fact they were probably similar to Croads in overall shape. Two plumage colour varieties were recorded, White and 'Ermine' (as Light Brahmas). Croad Langshans may have been used at a later stage in the development of Malines. Croads were not known in Europe until 1872, so could not have been involved at the early stages.

The previously existing Belgian breeds used in the mix probably included Silver Campines/Brakels (then not fully divided into two distinct breeds) and Coucou de Flandre/Vlaanderse Koekoek/Flemish Cuckoo. The latter are the most likely, and may be regarded as an 'original type' of Cuckoo Malines kept in the region before the arrival of the Asiatics allowed the development of a larger breed.

Like the Lincolnshire Buffs previously described, Malines were practical table birds bred to supply the Brussels markets, so it is not surprising that no one wrote a breed standard until 1895. Even then, the standard was not officially recognized until 15 June 1898. A specialist 'Club van de Mechelse' was formed on 25 March 1909. Paul Monseu was Chairman, A. Van Schelle was Vice-Chairman, E. Feyaeris was Secretary and A. Van den Kerkhoven was the 'penningmeester'.

Once they became show birds, the biggest birds usually won the top prizes. The more competitive breeders resorted to crossing with Belgian Game, the biggest chickens available. Although far too slow-growing to be of any use for meat production, some Belgian Game cocks eventually weigh over 6kg (13lb). This resulted in a pea-combed variety, often called 'Kalkenkop', or turkey-headed.

Malines are supported in their home country, but are rare elsewhere. They have been bred in modest numbers in the UK, sporadically, but have not become a permanent part of the fancy here. No references to Malines bantams have been found.

Description

General body shape is similar to Sussex, plus the lightly feathered shanks and feet. They are large birds, even pullets weigh 3kg (6½lb), cocks are 4.5–5kg (10–11lb). The main traditional colour variety is the Cuckoo. The barring is much clearer than that of Cuckoo Marans, but perhaps not quite as sharp as that of the best Scots Greys. White Malines were mentioned by Edward Brown so must have been well-established by 1905. Rarer varieties include: Black, Ermine (as Light Brahmas), Blue, Silver-cuckoo and Gold-cuckoo.

MANTES

Mantes, or Du Mantes, were developed by M. Voitellier, who lived in the town of Mantes, to the west of Paris. He first brought them to public attention in 1878. Some poultry experts have stated that one of the breeds he used to make them with was the Houdan, a reasonable assumption as Mantes have black-mottled plumage and beards (but not crests, and only four toes), but M. Voitellier always denied this. The mottled plumage probably really came from the Gournay, a light breed from Gournay-en-Bray in Normandy. Alternatively, both breeds may have come from a common ancestor. Other breeds that may have been used to make Mantes may have included La Bresse and some heavier breeds such as Dorkings or some now unknown types of Asiatics (Brahmas, Cochins or Shanghais). This happened before Faverolles were established as a recognized breed, but the birds used may have been 'prototype Faverolles'.

European Medium / Heavy Breeds

Cuckoo Malines, pair. Artist: Kurt Zander

BELOW: White Malines, pair. Artist: Kurt Zander

Current strains of Mantes are from a strain developed about 1960, as the originals died out, probably during the Second World War. Although some Mantes have no doubt been exported to other countries, they have never held anyone's interest for longer than a season or two.

Mantes are a medium-sized breed, ranging from 2kg (4½lb) pullets up to 3.5kg (7¾lb) adult cocks. They have single combs (upright on mates, flopping over on females) and well-developed beards. The mottled plumage is irregularly patterned, similar to that of early Houdans, and many utility strains of Houdans in France today.

NIEDERRHEINER

These were first developed by two German breeders Jacques Jobs and Friedrich Regenstein starting about 1939. They were trying to make a new, and very definitely German, meat breed. The breeds they used to make them were not very German however – Malines from Belgium, North Hollands from The Netherlands, Orpingtons from England and Plymouth Rocks from America. All of the traditional German breeds were small laying types, so there may have been an element of nationalism in the desire to create a German table breed.

The first colour variety made was unique to standard bred poultry, the 'Blauesperber' or 'Blaugesperbert', a Cuckoo (as Malines and North Hollands) modified by the Blue gene. Friedrich Regenstein must have been keeping track of the activities of the team of geneticists at Cambridge who were developing the Autosexing breeds in England, as the next two colours of Niederrheiners were Autosexers: Kennsperber and Gelbsperber. These two colours were recognized in 1943, evidence that at least some normal life was continuing during the war. Blue Niederrheiners were first seen in 1965.

Niederrheiners survived the war; they were then a practical meat breed during the 1950s until replaced by hybrid broilers in the 1960s. They also became a popular exhibition breed in Germany. At the time of writing, 2005, the author knew of only one flock of (Goldsperber) Niederrheiners in Britain.

Zwerg-Niederrheiner (Niederrheiner Bantam)

These were developed in both the DDR (East Germany) and FDR (West Germany) between 1950 and 1965. All available references state that they were made by selection of the smallest available large Niederrheiners only. If any crossing with pre-existing miniature heavy breeds was carried out (Orpington or Sussex Bantams perhaps?), they did not admit it. The miniature version of the breed is currently more popular in Germany than the large fowl.

Description

Niederrheiners are robust birds of good size, but not giants of the poultry world. Large fowl range from 2.5kg (5½lb) pullets up to 4kg (9lb) adult cocks. The equivalent range in the miniatures is from 900g (32oz) pullets up to 1,200g (42oz) adult cocks. Their backs appear short because they have saddle plumage that rises to the tail; although not to the extent seen on German Langshans. They have prominent, rounded breasts as should be expected on good table birds. Wings are carried 'well up'. Combs are single, of medium size, and straight. Lobes are red. Eye colour ranges from orange to brown.

Plumage colours (German names used throughout in this case, because some are unique to this breed):

Birkenfarbig: these are Silver-hackled Blacks. Many British fanciers would say they look just like Norfolk Greys except for the Niederrheiners' light shanks and feet

Blau-Birkenfarbig: as above, but an even blue instead of black

Blau: medium blue shade throughout, except for darker, shiny neck and saddles of males. The breast plumage should be even, without lacing, if possible

Blausperber Neiderrheiner, large male. Artist: August Rixen

Blausperber Neiderrheiner, large female. Artist: August Rixen

- Blausperber: a very attractive bluish-grey ground colour with dark cuckoo barring throughout, Gelbsperber: a buff Columbian with barring. The barring is naturally more distinct on the black parts. British poultry people have not seen this, or a similar, plumage pattern since the long extinct Brockbars, Buffbars and Golden Barred Plymouth Rocks. Gelbsperbers were made from Kennsperbers (*see below*), Buff Orpingtons and Light Sussex
- Kennsperber: these are Creles, broadly similar to Welbars in the UK. They differ in being more reddish, with less partridge markings. Kennsperbers were made from Blausperbers, Rhode Island Reds and Welsummers.

NOIRE DU BERRY

Berry-Bouy is a village to the west of the central French city of Bourges. The breed of chicken named after it was developed in the surrounding area about 1900. As the breed's name suggests, there is only one colour variety, Black. Black La Bresse were the traditional, high-quality but small, meat breed of eastern France. Fine for gourmets, but perhaps not economic for normal consumers. To increase size, they were crossed with Brahmas, Black Langshans and Black Orpingtons. They were officially recognized in 1912, but died out and had to be made again. When they first died out is unknown to the author, but considering the history of France from 1914 to 1945, it is not surprising. The second strain was made using Black La Bresse, Black Orpingtons and Noir de Challans (*see opposite*). The author doubts if this breed is bred anywhere outside France. No references to a bantam version have been found.

They are a typical medium-weight, single-combed, black-plumaged breed. Weights range from 2.3kg (5lb) pullets to 3kg (6½lb) adult cocks. As such, they are unlikely to attract much interest from poultry exhibitors outside France.

NOIRE DE CHALLANS

This is another single-combed, black-plumaged breed, very similar to the Noire du Berry described above. They were also made about 1900, also using Black Langshans and Black Orpingtons with local fowls. Challans is a town near the south-western coast of France, south-west of the city of Nantes. No references to a bantam version have been found.

NOORD-HOLLANDSE HOENDER (NETHERLANDS TYPE)

These originated about 1900 in the Noord-Holland province of The Netherlands, the area near the coast, north of Amsterdam. They were specifically bred near Zaanstad to supply the Amsterdam markets with table cockerels. Malines/Mechelse fowls from Belgium were tried, but they did not thrive in this cold, windswept area. However, when Malines were crossed with smaller, hardier local breeds, a successful compromise between size and survivability was found. Noord-Hollandse Hoenders were not standardized until 1930, as it took thirty years to stabilize their appearance. Much of the variation was in foot feathering, as Malines are feather-footed, and the local breeds they were crossed with were clean legged. Cuckoo Barred is the only standardized colour variety, but there were once Whites as well.

Noord-Hollandse Kriel (North Holland Bantam)

These were made about 1960 by Mr Meijer using several bantam breeds including British type Marans Bantams, Niederrheiners and Sussex. Like the breeds they were made from, Noord-Hollanse are quite large, necessary to retain a table breed type. They are generally very good miniatures of the large fowl, except for a tendency to have tails that are too large and too fanned.

Description

This is a very conventional-looking medium/heavy breed, deep-bodied with rounded breasts and a broad back. They have straight single combs, normal wattles, red lobes and orange-red eyes. When standing normally, their back line is nearly horizontal with fairly upright but short and compact tail. Their skin is white, necessary for a good table breed, and shanks/feet are white to light pearl-grey. Plumage is cuckoo barred that (as a guide for readers in the UK at least) is of a shade and clarity halfway between Marans and Scots Greys.

Standard weights of the miniature version are 900g (32oz) F and 1,000g (35oz) M. The large fowl range from 2.75kg (6lb) pullets up to 4kg (8½lb) adult cocks. Size is important for both.

Noord-Hollandse, Netherlands type, large male. Photo: John Tarren

European Medium / Heavy Breeds

NORTH HOLLAND BLUE (BRITISH TYPE)

Noord-Hollandse Hoenders were imported into the UK in 1934, shortly after Cuckoo Marans came in from France in 1929. The British have a tradition for not being very good at foreign languages, and poultry keepers, even the experts, were probably worse than average. Noord-Hollandse were still a variable new breed in 1934, as were Marans. Clean legs and feathered legs were no doubt seen on both. Thus, through ignorance, Marans were standardized with clean legs in the UK, and 'North Holland Blues' with feathered legs, both the opposite of the (correct) standards in their home countries. They were called 'Blues' because the dark grey/light grey barring was said to appear blue from a distance. This standard was passed by the PCGB on 17 July 1946. Goodson's NHB Farm Ltd. of Lenham, Kent, claimed to be the originators of NHB in Great Britain.

One of the initial breeders was Lv.Zwanenberg of Totteridge Lane, Totteridge, on the northern outskirts of London. North Holland Blues never made much impression as show birds, but were quite popular as table birds until they were replaced by hybrid broilers about 1960. British-type North Holland Blues only survived through the efforts of one man, Leslie ('Les') Charles Watkins Miles, and his back garden flock at Brigadier Hill, Enfield, near Totteridge. A founder member of the Rare Poultry Society in 1969, Les continued to keep North Hollands as long as his health permitted. He died, aged 92, in March 1992. During the 1970s and 1980s, Les probably knew the location of every North Holland Blue in the UK, making him the most efficient 'Breed Registrar' the RPS ever had. In 2002 it was estimated that there were about sixty breeding females in the UK, which, as the feather-legged NHB is a purely British type, is also the total world population. There has never been a bantam version.

They could be described as being the equivalent of Cuckoo Croad Langshans, and are similarly large birds, 2.7kg (6lb) pullets up to 4.8kg (10½lb) adult cocks. Females usually have

North Holland Blue, UK type, large male. Photo: John Tarren

North Holland Blue, UK type, large female. Photo: John Tarren

more shank feathering than males. In general colouring, males are lighter than females. According to 'Goodson's', up to 85 per cent of the chicks could be sexed at day old. This should not be taken to extremes, males should be barred throughout – not with white in wings or tail. Similarly, females should not be so dark that they lose clarity of barring.

NORFOLK GREY, ORIGINALLY 'BLACK MARIA'

Mr Fred W. Myhill of Norwich first exhibited his new 'Black Marias' at the 1920 London Dairy Show. He always insisted that the name should be pronounced Mar-e-a, and they were not named after police vans, but the nickname of a First World War military item. During the war 'Black Marias' were both a type of artillery shell and a black Sopwith triplane, the first to have the timing of the propeller blades synchronized with the machine gun.

In the 1923 *Poultry* (magazine) yearbook, Mr Myhill said 'its shape suggesting a cross between a Dorking and a Wyandotte'. Other sources said they were really bred from Birchen Grey Old English Game, Duckwing Leghorns, and some larger fowls such as utility Orpingtons or Langshans. Despite Mr Myhill's best efforts, and claims that they were selling well and were widely admired, the truth was less rosy.

About 1925 the name was changed to Norfolk Grey, which for some reason seemed to help their popularity. It is certainly true today (2005) that most of the limited number of hobbyists who keep them live in or near Norfolk. A Norfolk Grey Club was formed in 1927 or 1928, the secretary being Thomas Leyson of Stowmarket, Suffolk. Club Shows were held at Crystal Palace, with Norfolk Grey classes put on a limited number of other shows.

Things started to go wrong as soon as 1929/30, when the number of exhibitors and birds appearing at the shows began to decline after an all too brief bright start. Mr Leyson suggested one reason for the decline was:

The encouragement [by which he presumably meant show awards] of too small a type of bird. The Norfolk Grey is classed as a heavy breed by The Poultry Club, and the public is not used to seeing heavy breeds no larger than Old English Game, and it certainly will not have them. This type has crept in during the past three years, and it is very different from the original Norfolk Grey introduced by Mr Myhill. Moreover, this small Gamey type is nothing like as good a layer as the old type. I suggest a return to the larger bodied type, even though it may be at the expense of markings for a year or two.

This change was almost certainly the result of crossing with dark Grey Old English Game to improve plumage markings. Mr Myhill's originals laid a lot of large tinted or brown eggs, whereas the show winners laid fewer, smaller, off-white eggs; just like Old English Game.

There were never any really large displays of Norfolk Greys at the shows, the most found by the author being thirty-two birds, spread over three classes from eight exhibitors at their 1929 Club Show. Despite declining entries, the Norfolk Grey Club continued until it closed down for the Second World War.

Nothing seems to have been seen of Norfolk Greys until about 1970, which suggests that many Norfolk Greys around now are from a 'facsimile' strain made circa 1967–73, probably from Dark Grey OEG and Silver Sussex. It is possible that at least some can be traced back to the 1950s, and if so, they may trace back to a pre-war strain. No Norfolk Grey Bantams have been seen, although British Standards give suggested weights.

Description

Norfolk Greys are a very conventional shaped breed, so the two main features which breeders and judges will concentrate on are size and colour. Pullets should weigh at least 2.25kg (5lb) and cockerels 3.2kg (7lb), both gaining at least 400g (14oz) by their second

European Medium / Heavy Breeds

year. These weights are the same as those of Barnevelders, which should give an idea of the size required. Because Norfolk Greys, especially males, have larger tails than Barnevelders, a Norfolk Grey cock would appear generally bigger than a Barnevelder cock, if both were the same weight. Few Norfolk Greys seen at the shows since the 1970s have looked larger than Barnevelders. The breed standard, and photos of winners in the 1930s, have 'fairly upright and very active' carriage. Too many are rather too low and horizontal.

They have medium-sized, straight and upright single combs, small ear lobes and fairly long wattles. All head skin, including lobes, should be red. Pullets especially often have darkish pigmented faces, to be expected for a variety with the dark eyes and plumage pattern of Norfolk Greys. Shanks and feet are black (younger birds) or slate-black (older birds).

The main plumage colour is glossy black. Hackle feathers, just the neck of females, neck and back of males, should be silver-white with black centre striping. A tiny amount of silver lacing is allowed on the front upper neck, but should certainly not extend down to the breast as seen on Birchen Grey Game. As this is at the dark end of the spectrum of this plumage pattern, Silver Sussex (with almost all-over lacing) being the lightest version on standardized breeds, there is likely to be a problem getting really white, white parts. Most Norfolk Grey cocks are distinctly yellowish in their top colour.

Special Management Requirements

Shaded runs and the other methods used by those keeping white, or partly white plumage, to avoid 'brassiness' apply to Norfolk Greys, especially males. They need to be fit and the nutrition must be correct to avoid the plumage faults ('fretting') often associated with black feathering. Inbreeding may reduce size, so keep a large flock if possible, and keep in contact with other breeders to exchange breeding stock.

Norfolk Grey, male. Photo: John Tarren

Norfolk Grey, female. Photo: John Tarren

SCHWEIZERHUHN

This Swiss breed is closely related to the German Reichshuhner. They might have remained as simply a variant of them, but when Swiss breeders were planning to exhibit their strain at the 1921 World Poultry Congress at Den Haag (The Hague), The Netherlands, not long after the end of the First World War, 'Schweizerhuhn' would be better received than anything too German sounding. This may have been a good idea, but Schweizerhuhn have remained virtually unknown outside Switzerland.

There is only one colour variety, White. This is not surprising as apart from Reichshuhner, Schweizerhuhn were made from White Orpingtons and White Wyandottes. They have compact Wyandotte type rose combs and Orpington-type white skin, shanks and feet. Weights range from 2.4kg (5¼lb) pullets up to 3.5kg (7¾lb) adult cocks. When standing normally, they have the horizontal back line and low tail carriage of Reichshuhner, but with the shorter tails and heavier body one would expect from the Orpington and Wyandotte influence.

SUNDHEIMER

Sundheim is a village on the German side of the Rhine, opposite the (now) French city of Strasbourg. Sundheimer fowls are the German equivalent of British Light Sussex and French Ermine Faverolles. The process of crossing that led to their development started about 1875, and as might be expected, they were very variable to begin with; some had beards, some not; some single combs, others rose combs. By 1886, they had become stabilized, and were first exhibited. At the same time, a breed club was formed, the 'Sonderverein der Züchter des Sundheimer Huhnes'. The club had periods of inactivity, the present club dating back to 1978. Details of their development have not been found, but Cochins, Dorkings and Houdans were all mentioned. Light Brahmas, Light Sussex and Ermine Faverolles may have been involved.

Zwerg-Sundheimer (Sundheimer Bantam)

These are a post-war creation, first appearing about 1952, but only becoming really established in the 1980s. Undersized large Sundheimers were crossed with Light Sussex Bantams, Cochin Bantams and Faverolles Bantams.

Both large and bantam Sundheimers have a moderate following in Germany, but remain virtually unknown elsewhere.

Description

British, and many other, readers only need to imagine a Light Sussex with slightly feathered feet to understand what a Sundheimer looks like. They do not have quite the long, flat back that Sussex do, instead the saddle rises, somewhat in the style of Brahmas.

Large Sundheimers weigh from 2kg (4¼lb) (pullets) to 3.5kg (7¾lb) (adult cocks); the bantams 900g (32oz) (females) up to 1,100g (39oz) (males).

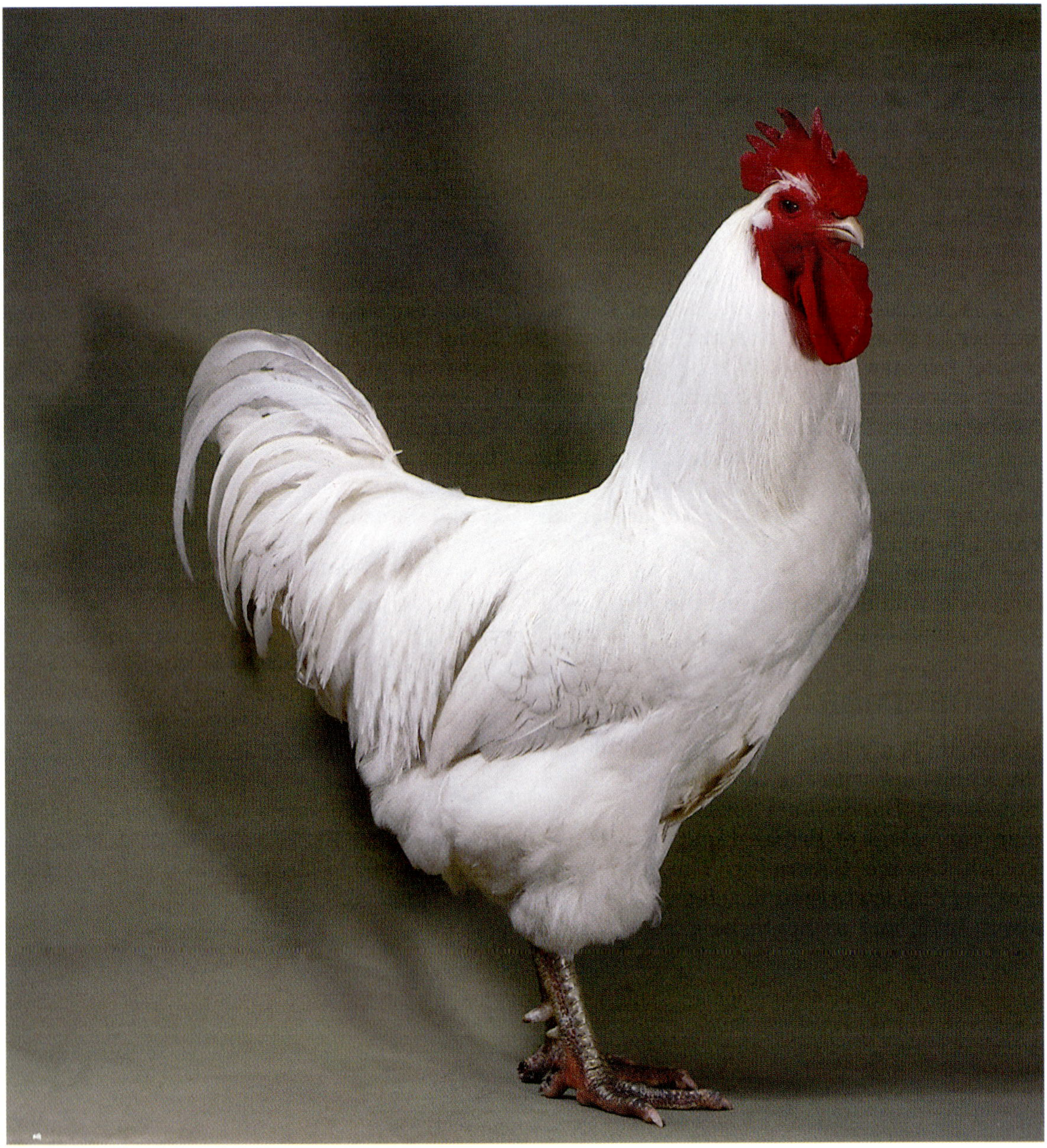

CHAPTER 12

American and Canadian Breeds

Large-scale egg and poultry meat production, a major industry all over the world now, was an American development. Until American business enterprise demonstrated the possibility of industrial-scale poultry keeping, it had been a minor part of rural economies. American companies also led the way in developing large incubators, refrigeration equipment and other machinery necessary for large-scale production. Most American poultry breeds were designed to suit market requirements, such as American consumer preference for yellow skinned table birds.

Although poultry shows were significant for American poultry farmers, they were never quite as dominant as in Britain in the main period of breed development from 1850 to 1940. For example, double mating was regarded as an undesirable complication in the USA, wheras leading British breeders regarded double mating as a useful way of keeping novice breeders 'in their place'. Laced Wyandottes were intended to be a 'gentleman farmer's' fowl, to provide eggs and meat, as well as looking good when strutting around in front of 'the big house'.

New Hampshires, Plymouth Rocks, Rhode Islands and Wyandottes have become major international breeds, and so too popular to be included in *Rare Poultry Breeds*. This book does not generally cover Game Fowl breeds, and is certainly not going to attempt to even list the many named strains of American Pit Game, a subject that has already filled at least two books. To give a taste of the subject, here are a few of the strain names: 'Alabama Roundheads', 'Arkansas Travelers', 'Georgia Shawlnecks' and 'Grey Tormentors'. As the names suggest, most of the breeders were in the once Confederate southern states.

The breeds are covered in this chapter in approximately chronological order:

Dominique: standardized 1874, but known to exist before 1800
Java: standardized 1883
Buckeye: standardized 1904
Chantecler: standardized 1921
Jersey Giant: standardized 1922
Lamona: standardized 1933
Holland: standardized 1949
Delaware: standardized 1952.

Extinct breeds:

Albions/Bristol White/White Bristol
American Sebright
Birmingham White/White Birmingham
Bristol County Fowl
Bucks County Fowl
Danvers Whites/White Danvers
Eureka
Golden Buff
Hambletonian
Jersey Blue (not the same as a Blue Jersey Giant)
Oneida County Fowl
Bennett's first 'Plymouth Rock' (not the same as the main Plymouth Rocks)
Seabright Cochin

Sherwood
White Wonder Fowl
Winnibago.

DOMINIQUE

Despite thorough investigations by Mark A. Fields, author of *The American Dominique* (1997) the exact origin of America's oldest breed remains unclear. His book was prepared using all the archive collections of the club membership. The first poultry show in America was held on 15–16 November 1849 at Boston. There were four exhibitors with Dominiques at this event, along with some other early American breeds (Bucks County Fowls, Jersey Blues, Javas and Dr Bennett's first Plymouth Rocks).

Dr John C. Bennett, organizer of the 1849 Show and author of *The Poultry Book* (1850), said, 'I have never witnessed the least variation in their appearance, for the last thirty years.' This would be referring to about 1820, and if they were already breeding true then, we can assume they probably go back to sometime near 1800. Dr Bennett thought they were called Dominiques because they originated from the island of Dominica, and were brought to the American mainland by French settlers.

It is possible that at least the name came from Dominica, or possibly Santo Domingo, the main port on what is now the Dominican Republic that was then 'Hispaniola', and was at various times under French rule, Spanish rule, or independent (pirates' rule?). Considering the type of societies in the Caribbean during the eighteenth and early nineteenth centuries, the only chickens likely to have come from there to contribute to the Dominque breed would have been Cuckoo and Crele-coloured Pit Game. The name 'Dominique', or variations such as Dominic and Dominick were used to simply describe the barred plumage pattern. It has also been suggested that 'domino' might be the source, suggesting a combination of black and white.

All of America's domestic chickens were brought over by settlers who would naturally bring their own country's breeds: the British bringing Cuckoo Dorkings, Scots Greys and Scots Dumpies; French settlers could have brought Coucou de Rennes; and people from the rest of Europe would have brought a whole range of breeds, including Brakels and Hamburghs. None of these breeds have the Dominique's yellow skin and feet, this gene probably coming from willow or yellow-legged Game breeds from England, Spain and South-East Asia Yellow-skinned table birds soon became the favourite of American consumers.

It is quite possible that 'Dominiques' developed independently in two places from two separate (or at least mostly separate) mixes of breeds. Up the east coast, in the more Puritan British Thirteen Colonies as they were first established, Dorkings, Scots Greys and Hamburgh-type birds would have been brought over. The deep south was certainly not a Puritan area, and cockfighting was so popular that almost every flock of chickens would have had some Game Fowl ancestry. *Johnson's History of Game Strains* gives an example of the relationship between Gamecocks, chickens in general and Dominiques (or at least 'Dominique'-coloured fowls):

> In the 1830s there was being bred on Rabbit Island, near New Orleans. La., a yard of imported English-Spanish fowl. From this yard came an off coloured stag, a speckled yellow, blue and white with rose comb and yellow legs. At age of three years he had never been bred or fought but the owner decided to carry him to New Orleans and fight him. He was laughed at and called a barn yard dunghill but won several quick fights.

This passage goes on to describe how this cock was sold and bred with proper Game hens to establish a strain of 'Kentucky Dominiques' Game Fowl. The term 'Dominique' was applied to cover both pure black and white barring and the Crele pattern, barring over the Black-Red/Partridge pattern.

Although recognized as the first widely established American poultry breed, Dominiques soon faded into obscurity. They were mainly supplanted by the new Barred Plymouth Rocks that were made by crossing Dominiques with recently imported Asiatic fowls. The early Brahmas, Cochins, Javas and Langshans were not as feathery as they would become, and their sheer size attracted a lot of interest. Several breeders had been trying these crosses for some years, with the first birds shown as 'Barred Plymouth Rocks' appearing in March 1869 at a show in Worcester, Massachusetts by Mr D.A. Upham of Wilsonville, Connecticut. The first edition of *The American Standard of Perfection* in 1874 included both Dominiques and Barred Plymouth Rocks. These standards settled two differences between the breeds, size and combs. Rocks had been intentionally made larger by the Asiatic crossing, although this still left some scope for debate about the desirable size of Dominiques. In short, everyone agreed that a Dominique should be smaller than a Rock, but how much smaller? Both single and rosecombs had been seen on both Dominiques and Barred Plymouth Rocks, and even a few pea-combed Barred Rocks appeared in the 1880s. This potential confusion was settled by establishing rosecombs for Dominiques and single combs for Rocks. The barring was similar on both breeds in the early years, relatively coarse, but by about 1900 a clear difference could be seen. Barred Plymouth Rocks were being shown in considerable numbers, requiring sharply defined markings to stand any chance of winning. When the judges were looking tor 'Best American' or 'Show Champion', the winning Barred Rock was bound to look more impressive than the best Dominique. Specialist table bird producers, especially those aiming for the larger, luxury end of the market, would also prefer Rocks to Dominiques. However, there were enough practical poultry keepers who appreciated the qualities of the old breed to keep them alive until wider interest revived. In a general farm, smallholding or garden situation, Dominiques are arguably a much more useful breed. Being smaller, and probably more active birds, they need less food and if allowed to roam freely could probably find a lot of it themselves. They are also much quicker feathering than finely barred Plymouth Rocks. In the cold winters of the northern states of the USA compact rosecombs are less likely to suffer from frostbite than single combed birds.

The current Dominique Club of America was formed in 1973 and with the present general interest in the conservation of old breeds is very successful. It is a pity that previous generations of Dominique breeders were not as well organized or motivated as the current enthusiasts. There have been several breed clubs over the years but none of them seem to have lasted very long. Mark Fields, Secretary/Treasurer of the current club, lists these previous organizations:

A very early (circa 1827) Dominique Club at Salem, Massachusetts. No details known, but it may have held informal competitions as British breeders did at the same period with Hamburghs, Sebrights and Spanish.

The American Dominique Club, short-lived, circa 1898.

National American Dominique Club, short-lived, circa 1910–12. In 1912 it had 103 members. The President was Dr. H.W. Skerritt, Utica, NY and the Secretary A.Q. Carter, Freeport, Maine.

Dominique Club of America, 1924 only, then changed its name to National American Dominique Club after contacting and combining with remaining members of the above club that had been inactive for some years. It is not known for how long this club remained active, but it seems to have closed down before the Second World War.

American and Canadian Breeds

This chequered history of clubs forming and then closing down cannot have helped the Dominique's popularity during a period when other breeds had powerful clubs with huge memberships. Mark Fields' book gives biographical details of some of the leading breeders. It may now never be known if there were any personality clashes or reclusive breeders that reduced the chance of a strong breed club lasting straight through from the nineteenth century.

Despite the problems with the clubs, from 1900 until the Depression of the late 1920s there was at least a great improvement in the uniformity and quality of the limited number of Dominique flocks. A.Q. Carter, the first Secretary of the National American Dominique Club (1910 to 1918?) was the owner of the birds that appeared in the APA Standards for many years as the ideal for everyone else to aim for.

Dominiques survived from 1930 through to 1970 largely due to the perseverance of two dedicated breeders, Robert F. Henderson (Zanesville, Ohio) and Edward H. Uber (Chicora, Pennsylvania) who were later joined by J. Cari Gallaher (Ellwood City, Pennsylvania). Several breeders took an interest in 'America's oldest breed' during the 1970s, as fully described in Mr Fields' book.

'Dominikaner' in Germany

Apart from these American enthusiasts, the only other significant group of breeders has been in Germany. They were first imported to Germany in 1880, and survived through to the present despite the two world wars. Some German breeders used Dominikaner to produce Barred Wyandottes by 1908. This has probably influenced German Dominikaner breeders to favour a slightly smaller-bodied, longer-tailed type than American Dominique fanciers. They needed to be sure that no one could possibly accuse a Dominikaner of just being a poor Wyandotte or Dresdner. Sumatras were crossed with American Dominiques to obtain a clearly different tighter-feathered, longer-tailed type. Many of the old drawings and descriptions favour the smaller body size, tighter plumage and large fanned tails of the German Dominikaner, but the Sumatra influence has also brought small combs and low tail carriage, which is a clear departure from the historic type.

Dominique Bantam

Dominique bantams have been made, died out, and re-made, several times since the 1870s. Mark Fields noted these appearances:

Pennsylvania State Show, 1873. *Poultry Monthly*, 1889, included a picture of a Dominique Bantam.
Hagerstown Show, Maryland, 1914 included Dominique Bantams entered by Ore Lake Farm.
American Bantam Association, 1935 *Yearbook*, G.F. Wright noted them as recent creations.
American Bantam Association, 1968 (Yearbook?), G. Fitterer describes the making of a strain.
Mr Glenn F. Perry of Franklin, Ohio had a strain through much of the 1970s. Other breeders at this time were Ralph Brazelton, Elva Hemphill, Christina Kiser and Samuel LeCotes. They are mainly the result of crosses between undersized large Dominiques and White Rock Bantams.

Currently (2005) Dominique Bantams are bred by keen Dominique Club members, but remain very rare.

German Zwerg-Dominikaner

As with the large version, German bantam (Zwerg = dwarf) breeders wished to avoid any suggestion of Plymouth Rock, Wyandotte or Dresdner type. Wandelt and Wolters (*Handbuch der Zwerghuhnrassen*, 1998) record two breeders, Fritz Pohle and Erich Kowert, using large Dominikaner (already containing Sumatra characteristics), and bantam Barred Italiener, Plymouth Rocks, Rheinlanders and Sperbers, plus Silver Kraienkoppe Bantams. The result is an elegant breed, but still with rather small combs and low carried tails to be accurate miniatures of historic Dominiques.

Dominiques, Large and Bantam, in Britain

Since 2000 very limited numbers of both sizes have been imported from both Germany and America. At the time of writing, 2005, it is not clear what direction to go from these two different forms of one (nominal) breed. The author suggests that the strains be crossed to obtain birds with the body and tail size of the German birds, with the combs, plumage pattern and tail carriage of the American strains. Such a type would seem to closely resemble the old pictures of Dominiques and clearly avoid any suggestion of Wyandotte type.

Description

The basic requirements of a Dominique are simple: rose comb, red ear lobes, barred plumage (clear, but not as refined as on Barred Plymouth Rocks), yellow skin and legs, and of medium size, smaller than Barred Rocks. The devil is in the detail.

Comb: Dominique style rose combs are relatively large with well-defined 'work' on the top surface and a leader that sweeps well up from the skull, nearly or slightly above the top surface of the comb. Breeders should be aware, especially on males, that cockerels with correct size combs may have over-sized and misshapen combs when they are older. In contrast, cockerels with rather small combs, may make excellent show birds when three or more years old.

The combs should be moderately broad, but not so broad as to droop over the eyes. They should also have reasonably well-developed 'workings' on the top surface. That is, more 'work' than on most Wyandotte combs, but less than on a Derbyshire Redcap. The rear part of the comb should have a definite curve into the leader, not a gradual tapering as on Wyandotte combs. The leader should be round in cross-section, not blade-shaped.

Standard Weights: N.B. Only American and German weights are given, as these are the two key countries for Dominiques/Dominikaner:

- USA – Large: cock 7lb (3kg), cockerel 6lb (2.7kg), hen 5lb (2.25kg), pullet 4lb (1.8kg)
- USA – Bantam: cock 28oz (790g), Cockerel 26oz (740g), hen 24oz (680g), pullet 22oz (620g)
- Germany – Large fowl: male 2 to 2.5kg (4½ to 5½lb), female 1.75 to 2.25kg (4 to 5lb)
- Germany – Bantam: male 900g (32oz), female 700g (25oz).

Beak, Skin and Shank/Foot Colour: All these parts should be bright yellow, which can fade to cream if Dominiques are not given access to grassed runs and/or have some maize in their diet. Some hens may still fade towards the end of a laying year, but the colour should brighten again during the moult. A little black shading on the shanks and toes or black markings on the beak are sometimes seen on all Barred breeds. If not too heavy, it should not be regarded as a major fault.

Plumage Pattern: American Dominique specialists have long been keen to avoid barring, which is fine and regular like that of Plymouth Rocks. Those new to the breed should study all available literature on the breed and all available live Dominiques to fully understand the ideal. The overall shade should be quite bright, avoiding the dull, dark shade British readers see on Cuckoo Marans hens. Old cocks sometimes moult with nearly all white wing and tail feathers. They are probably safe to breed from if they were soundly marked in their first two adult years. Double Mating has been used by some specialist breeders.

JAVA

American poultry keepers in the nineteenth century gave the name 'Java' to almost any

American and Canadian Breeds

Dominiques and Mottled Javas. Artist: Carl Fielder. One of a set of 75 (+?) postcards published by Albert Hoffman, Magdeburg, 1914.

black, or mostly black-plumaged fowl imported from the Far East. Some were Malayoid Game-type birds, others more heavily feathered, probably similar to Croad Langshans in general appearance. From this mixture a specific American breed called the 'Java' was standardized by the APA in 1883. By this time a second colour variety, the Mottled, had already been established. We will probably never really know if American Javas had any connection with the island of Java, but at least their history in America was recorded.

Dr John Bennett's *Poultry Book* (1850) gives an early illustration and description of 'Great Java Fowl' that are broadly similar to the type standardized over thirty years later: large birds with single combs, non-feathered legs and with a plumage formation that is clearly a utility chicken, not a fighting cock. They were the property of Mr Sidney Packard of East Bridgewater, with similar birds also said to be kept by Mr Charles Burton of Plymouth, Massachusetts.

The stock on which the 1883 standard was based had a very specific origin according to a story originally published in *American Farmer* (1882) and retold in the American edition (1905) of Harrison Weir's *Poultry Book*. This story was told by Mr J.Y. Bicknell, of New York, a well-known Java breeder during the 1870s and 1980s.

A doctor in Missouri had a much admired flock of 'Javas' about 1850. He would not sell stock to any of his neighbours, so his coachman 'borrowed' three eggs that were hatched by his friends. In about 1857 the family moved back east to Duchess County, New York, and in 1867 moved again, to Orleans County, New York. They bred them in large numbers, so spreading them around New York State by 1870.

Mr Bicknell also described the origin of the Mottled variety. He first bred them by crossing a Black Java cock with a large white hen of unknown origin. The result was a strain of birds with a random mix of black and white plumage as also seen on Exchequer Leghorns. This was clearly required in the 1890 edition of *The American Standard of Perfection* ('... white and black feathers evenly intermixed'.), and a photograph of a cock like this appears in the 1905, American, edition of Weir's *Poultry Book*. By the 1920s there were probably not many Mottled Javas in existence, but Lamon and Slocum's book (*The Mating and Breeding of Poultry*, 1927) indicates there was a shift underway towards black feathers with white tips, as required now.

Javas were only used commercially in significant numbers until the late 1880s. They were briefly a major breed for table bird production for New York City and urban parts of New Jersey. Between 1865 and 1890, Javas were also used by breeders in the creation of Plymouth Rocks and Jersey Giants, the breeds that would replace them as commercial table chickens.

The decline was so severe that John H. Robinson, in 'Popular Breeds of Domestic Poultry' (pub.: *Reliable Poultry Journal*, 1924), doubted the authenticity of the 'Javas' entered in the shows at the time. Javas did survive however, the conservation effort being outlined by Pete Malmburg of LaFox, Illinois in the June 1998 issue of *'Poultry Press*. As with several other breeds, there have been a succesion of dedicated fanciers who kept them going until interest in the conservation of old breeds grew in the 1970s.

Pete Malmburg stated that the only major breeder of Javas through the 1950s was Duane Urch, then proprietor of 'Turnland Poultry'. He was still listed as one of only four Java breeders in the *Register of Poultry Genetic Stocks in North America* (circa 1975), at that time living at Owatonna, Minnesota. There have been so few other Java breeders recorded through to the 1980s, that they all deserve a mention: Acre Bird Farm (Wisconsin), Robert Pryor (Indiana), Clay Steinberger (Connecticut), Ted Stevemer (Minnesota) and Howard Tallman (Florida). Pete Malmburg first added them to his Garfield Farm Museum (Illinois) collection in 1986. By the time of his 1998 article he had contacted all other known Java breeders and had his flock DNA tested to establish their distinct identity from Jersey Giants.

Java Bantam

These were admitted to American Standards in 1960, but have never attracted much interest. They are likely to remain as a secondary interest for large Java enthusiasts in America and Canada only.

Description

Standard Weights (APA.): large: cock 9½lb (4.3kg), cockerel 8lb (3.6kg), hen 7½lb (3.4kg), pullet 6½lb (3kg). Bantam: cock 36oz (1020g), cockerel 32oz (900g), hen 32oz (900g), pullet 28oz (790g).

The standard weights clearly indicate these were intended to be useful general farm fowls, with perhaps more emphasis on meat than eggs. Their general appearance is similar to American-type Plymouth Rocks or Rhode Island Reds, differing by a slightly higher carriage, with the back line sloping instead of horizontal when they are standing normally. They also have larger and more spread tails than either of these two breeds. As they are very rare, care should be taken to ensure they remain large enough as inbreeding may reduce size.

They have moderately small, straight, single combs, red lobes and their eyes are dark brown on Blacks, reddish bay on Mottleds. Their shanks and toes have black epidermal scales with an undertayer (dermal cells) of yellow, darkest on young Blacks, lightest on older Mottleds. The yellow undertone is vital, indicating the yellow skin required on all American table chickens. Black Javas, and the black parts of Mottled Javas, should have a brilliant green sheen.

BUCKEYE

Buckeyes are similar to Rhode Island Reds in general appearance, differing by having pea combs and being of a slightly more muscular build. Indeed, they were first proposed for recognition by the APA as 'Pea Comb Rhode Island Reds', but this was rejected, so the alternative name was adopted by the originator, Mrs Metcalf, of Warren, Ohio. 'Buckeye State' is a nickname for Ohio. It is derived from the dark-brown colour of nuts from the horse chesnut tree (*Aesculus glabra*) which is common in Ohio, and was widely used by the early pioneers.

A.P.A. Standards mention Dark Cornish (= English Indian Game), Black-Red Game, Buff Cochins and Barred Plymouth Rocks as being used to make Buckeyes. Readers should remember that birds with these names in rural Ohio, circa 1890 to 1905, were probably rather different from present-day (2005) exhibition birds of the same names. In particular, the Cornish would have had much longer legs than present exhibition Cornish/Indians, and the Buff Cochins' were probably much tighter in feather, and with much less foot feathering, than exhibition Cochins.

Buckeyes were overshadowed by other more successful breeds both commercially and in the show halls, despite the efforts of the American Buckeye Club, formed in 1905, that had seventy-five paid-up members in 1912. Then the President of the club was the creator of the breed, Mrs Nettie Metcalf, who by this time had moved to Inglewood, California. The club was first named the 'National Red Feather Club', which had changed in 1911.

Five breeders were listed in the 1975 Register of Poultry Genetic Stocks in North America: Stromberg's (Iowa), George Pofahl (Minnesota), Duane Urch (Minnesota), Tier Haven (Pennsylvania) and Acre Bird Farm (Wisconsin). The 1998 APA–ABA National Show at Columbus, Ohio, had eighteen large Buckeyes, with six exhibitors named among the prize-winners. In 2005 there was one known flock of Buckeyes in the UK (Derbyshire).

Buckeye Bantam

These were admitted to the American Standards in 1960. Two strain originators, J.C. Kriner and S.B. Clem were mentioned by Fred P. Jeffrey in the early 1970s. Two exhibitors entered three birds each at the 1998 Ohio National, indicating that they have not attracted much interest.

Description

Standard Weights (APA): large: cock 9lb (4.1kg), cockerel 8lb (3.6kg), hen 6½lb (3kg), pullet 5½lb (2.5kg). Bantam: cock 34oz (950g), cockerel 30oz (850g), hen 28oz (800g), pullet 26oz (750g).

The overall appearance is very similar to Rhode Island Reds, except for the pea comb and slightly higher carriage when standing

Buckeye, large male. Photo: John Tarren

normally. Plumage should be a little tighter than on American-type Rhode Islands.

CHANTECLER

This is the only standard poultry breed created in Canada. They were the idea of Brother Wilfred Chateaine, a monk, who started his breeding operations in 1908. A compact cushion or walnut comb and very small wattles were key features, essential for cold winter nights before large-scale controlled environment poultry housing developed. In more traditional-scale poultry housing, the points of single combs could be badly frost-bitten.

The original white variety was the product of a mixture of: Dark Cornish (Indian Game), Rosecombed White Leghorns, Rhode Island Reds, White Wyandottes and White Plymouth Rocks, plus according to another account White Cornish. Brother Wilfred's project was taken further by the Oka Agricultural Institute, Quebec. The Oka Institute launched Chanteclers in 1918, and they were admitted to the APA Standards in 1921.

Partridge Chanteclers were created in Edmonton, Alberta, and standardized in 1935. They were made from Dark Cornish/Indians, Partridge Wyandottes, Rosecombed Brown Leghorns and Partridge Cochins.

They have survived, but are very rare. At the 1998 Ohio National there were just four large White Chanteclers from two exhibitors, and no Partridge Chanteclers at all.

Chantecler Bantam

White and Partridge Chantecler bantams were admitted to the APA Standards in 1960. At the 1998 Ohio National, one of the exhibitors of large Whites, Adrienne Blankenship, also entered one Partridge Chantecler Bantam.

Description

Standard weights (APA): large: cock 8½lb (3.8kg), cockerel 7½lb (3.4kg), hen 6½lb (3kg), pullet 5½lb (2.5kg). Bantam: cock 34oz (950g), cockerel 30oz (850g), hen 30oz (850g), pullet 26oz (750g).

The cushion comb gives Chanteclers a slightly gamey appearance, but they were intended to survive Canadian winters, so the body plumage should be dense and plentiful, but held fairly tightly, not loose like exhibition Cochins or Orpingtons. The correct colour and markings of Partridge Chanteclers is the same as American Standard Partridge Wyandottes, considerably darker than British and German Wyandottes.

JERSEY GIANT

The main colour variety of these is black. In addition to being (as one would expect) 'Black Jersey Giants', they have also been called 'Jersey Black Giants' and 'Blacks' Jersey Giants, the last deriving from their main originators, brothers John and Thomas Black who lived near Jobstown, New Jersey. Jersey Giants are really a further development of Black Javas, the main breed used by the Black Brothers. Increased size and hybrid vigour were obtained by crossing with Black Langshans, Indian Game and others. Breeding started sometime during the 1870s, when there were also still some utility strains of Brahmas in the eastern States. The Black Brothers and others in New Jersey, New York State and Pennsylvania were practical men whose main business was the production of large capons for restaurants in Boston, Philadelphia and New York City. Dark or black-plumaged birds were thought to be more disease-resistant and generally hardier than white-plumaged ones and they looked attractive when hung up partly plucked, common practice in the wholesale markets. The body was clean plucked, but the neck, wings, tail and hock joints were left feathered. Also, the birds were not eviscerated before going to market. However, Black Jersey Giants were not admitted to the American Standards until 1922, long after the Black Brothers' heyday. The American Jersey Black Giant Club had been formed in May 1921. U.L. Meloney is credited as being the leading breeder who made a uniform breed, one worthy of

Black Jersey Giant, large male. Photo: John Tarren

Black Jersey Giant, large female. Photo: John Tarren

standardization, from the more variable earlier stock. White Jersey Giants were added to the American Standards in 1947.

They were soon exported from America, the first British importers being Captain and Mrs Jones who took delivery of a trio in August 1921. They also quickly formed a breed club here for Giants. By 1924 Jersey Giants had appeared in Belgium and The Netherlands, but such big birds were clearly not suitable for the Great Depression years, and their yellow skin did not suit British or European market requirements.

The white variety was also imported to Britain, E. Hill of Ash Villa, Corsham, Wiltshire being the one importer recorded. A separate British Jersey White Giant Club had a brief existence from 1932 to 1936, reaching seventy-five members at its peak. Blue Jersey Giants seem to be a recent addition, possibly only dating back to the 1970s. Although they are probably very similar to the extinct Jersey Blue breed, the gap of almost a century means they have no direct connection.

Jersey Giant Bantam
Black and White Jersey Giant Bantams were admitted to the American Standards in 1960, but they have never attracted much interest. The 1975 Registry of Poultry Genetic Stocks in North America lists just one breeder, and the author could not find anyone else in ABA Yearbooks or American poultry magazines. As the main feature of the breed is its great size, there is not much point in a bantam version.

Description
Standard weights (APA and PCGB): large: cock 13lb (5.9kg), cockerel 11lb (5kg), hen 10lb (4.5kg), pullet 8lb (3.6kg). Bantams weights are rather academic, and the weights given in British Standards are much higher than those in the American Standards.

Most show judges, especially those who have not closely studied Jersey Giants, will expect them to be really giant. So the biggest bird usually wins the class. When viewed from the side, Jersey Giants should look long-bodied, with a horizontal back line. The colour of their shanks and toes is critical. On Blacks they are black with yellow soles on young birds, with the black fading to 'willow' (greenish) on older birds. Willow is the correct foot colour on Blues and Whites of all ages. Straight single combs, red ear lobes and dark brown eyes are required on all varieties.

Special Management Requirements
Because size is so important in show birds, nutrition is the key factor in rearing them.

Blue Jersey Giant, large male. Photo: John Tarren

Blue Jersey Giant, large female. Photo: John Tarren

Despite the cost, stick to commercial rations, because grain reduces the protein levels. Some breeders use broiler rations instead of the type intended for layer pullets used by most fanciers. Sensible breeders should make sensible decisions about the number of young Giants they can afford to feed to maturity. Even allowing for some cockerels being killed for the table before reaching maturity, breeders must remember that a batch of young Giants will cost a lot to feed and need a lot of house space. Houses should be regularly cleaned out as the young birds are best kept sleeping on the floor. Early introduction of perches will probably result in dented and bent breast bones.

LAMONA

Harry M. Lamon, Senior Poultryman at the United States Agricultural Research Center at Beltsville, Maryland, started the breeding programme that eventually produced the Lamona in 1912, but they were not perfected and admitted to the APA Standards until 1933. The three main breeds used to make Lamonas were White Leghorns, White Plymouth Rocks and Silver Grey Dorkings. The result was a white-plummaged bird with the

White Jersey Giant, large male. Photo: John Tarren

American and Canadian Breeds

long, low carriage typical of Dorkings. During the 1930s they were promoted by two breeders, Mr S.E. Raymond of Ohio and the Osborne Poultry Farm, Holland, Michigan, but they never really caught on. Lamonas were never intended to be an exhibition breed, and although they are still featured in American Standards, the author does not know if any still exist.

HOLLAND

This is another commercial breed that was replaced by hybrids and never attracted much interest among fanciers. The originators were Frank Gloeckl and Ross Salmon of Rucker's Imperial Breeding Farms, who started their breeding programmes in 1934. Hollands were standardized in 1849, in two colour varieties, Barred and White.

Breeding started in 1934 from some imported White Noord Hollandse (hence the name 'Holland'), Lamonas, White Leghorns, New Hampshires and Rhode Island Reds to make White Hollands. At the same time, a different mix was used to make Barred Hollands: Australorps, Brown Leghorns, White Leghorns and Barred Plymouth Rocks. This resulted in a breed that resembled a slimmed-down version of Plymouth Rocks, with weights ranging from 2.5kg (5½lb) pullets up to 4kg (8¾lb) adult cocks.

According to Craig T. Russell (SPPA President), writing in the October 189S issue of *Poultry Press*, White Hollands were believed to have died out about 1980, but Barred Hollands were still in existence, it might be possible to re-create Whites by crossing a suitable white breed with Barreds, but very few fanciers, and only those in America, are likely to be interested.

Holland Bantams were admitted to APA Standards in 1960, but it is doubtful if they ever really existed. The standard just gives suggested bantam weights, with shape and colour to be as for large Hollands.

DELAWARE

Mr George Ellis of Ocean View, Delaware, assisted by Dr Edmund Hoffman started a broiler chicken breeding programme in 1940 that eventually fed to the Delaware breed being accepted into American Standards in 1952. They started by crossing light-coloured Barred Plymouth Rock cockerels with New Hampshire Red pullets. The finished Delaware had mostly white plumage With black neck, wing and tail markings (as Light Sussex), but in this case broken up by the barring gene. This is a unique pattern, so Delawares have attracted enough interest among fanciers to survive through to the present day. Cecil Moore of Texas made the Delaware Bantam, accepted by the APA in 1860, but the author is not sure if these have also survived.

Delaware, large male. Photo: Hans Schippers

EXTINCT AMERICAN BREEDS

Albion/Bristol White/White Bristol
These were described by William E. Shedd, of Waltham, Massachusetts in an 1874 issue of the *Fanciers' Journal*. Mr Shedd had been advertising them as 'Albions', but 'Bristol White' or 'White Bristol' were the usual names for these birds that had been bred in Bristol County, Massachusetts since the 1850s. They were said to have started from a cross between a White Dorking cock and a White Asiatic (Cochin?) hen, followed by the use of another white cock of unknown origin. When White Plymouth Rocks were standardized in 1888, all those Bristol Whites that would pass as a White Rock were renamed as such.

American Seabright
(*N.B. This was the spelling used of Sebright.*)
Prototype Silver Laced Wyandottes. Bred by Mr L Whittaker of Michigan. They had rose-combs, laced plumage, rounded bodies and were clean – that is, not feather-legged; i.e. the same as the eventual Wyandottes. Thus Mr Whittaker's birds thrived, but his name for them did not.

Bristol County Fowl
Birds of this name were exhibited at the famous first Boston Poultry Show in 1849. As the name clearly indicates, these were also from Bristol County, Massachusetts. No detailed description of these birds has been found beyond the fact that they were generally red or buff in colour; no doubt bred from crosses of Partridge and Buff Cochins/Shanghais with farmyard hens. When Rhode Island Reds appeared in the 1890s, any surviving Bristol County Fowls were renamed as such.

Birmingham White/White Birmingham
No details of their origin were found, but D.A. Upham and I.K. Felch were said to have used them, with other breeds, to make Barred Plymouth Rocks. White Plymouth Rocks were standardized in 1888, and Birmingham Whites were merged with them. Things were nearly very different, as some breeders tried to keep the name 'Plymouth Rock' just for the original Barreds.

Bucks County Fowl
These originated in Bucks County, Pennsylvania, about 1800. Originally they were simply large farm fowls bred by crossing at first Malays and later Shanghais with local farmyard hens. By the 1870s, 'Yellow', 'Buff' or 'Straw Color' predominated. When Buff Plymouth Rocks were standardized in 1894, the remaining Bucks County Fowls were renamed.

Danvers White/White Danvers
These were named after the town of Danvers, Massachusetts about 1850; and were probably bred from crosses of Dominiques and White Shanghais/Cochins. Classes were provided for them at poultry shows in the New England states until the 1870s. Later, they were merged into White Plymouth Rocks and White Wyandottes, as both single and rose combs had existed.

Eureka
This was one of the prototype Silver Laced Wyandottes. Bred by a Mr Kidder, they had pea combs and feathered legs. Dark Brahmas, Silver Spangled Hamburghs (and?) were used to make them.

Golden Buff
These were bred by Dr N.B. Aldrich, of Fall River, Massachusetts in the 1880s, and first shown as 'Golden Buffs' in 1890. They were soon absorbed into Buff Plymouth Rocks, Rhode Island Reds and possibly also Buff Wyandottes, the initial birds being rather variable.

Hambletonian
One of the prototype Silver Laced Wyandottes. This was the name favoured by Mr

Bucks County Fowls, pair. From C. N. Bement's book, The American Poulterer's Companion, *1844. Artist unknown (from Fred Hams' collection)*

I.K. Felch when he wrote a provisional standard for what would eventually become the Wyandotte.

Jersey Blue

These were produced in the state of New Jersey, initially about 1815–20, from Malay-type birds crossed with local farmyard hens. They were fully described in the 1890 edition of the APA Standards that are summarized here:

Weights: cock –10lb (4.5kg), cockerel – 7lb (3kg), hen – 8lb (3.6kg), pullet – 5lb (2.2kg). Comb: single, medium-size, straight. Eyes: dark bay. Lobes: Red. Plumage: dark glossy blue hackles, remainder blue with dark lacing (as Andalusians). Tail: short, carried well up. Shanks and toes: dark slate, white claws, no feathering.

These details and the full standard describes birds were probably very similar to the later Blue Jersey Giants, although they were believed not to be directly related.

Oneida County Fowl

Another prototype Silver Laced Wyandotte, they were developed by Mr Payne of Binghampton, New York and are known to have existed in Oneida County, New York, as early as 1866.

The First Plymouth Rocks

These were described by Dr J.C. Bennett of Plymouth, Massachusetts in the *Boston Cultivator* of 25 August 1849. He fully described the process, bringing upon himself criticism that they were nothing but mongrels from other poultry breeders.

Dr Bennett wrote in 1850: 'I have given this name to a very extra breed of fowls, which I produced by crossing a cockerel of Baylies' importation of Cochin China with a hen, a cross between the Fawn-colored Dorking, the Great Malay, and the Wild Indian.'

'Baylies' importation Cochin Chinas' (Alfred Baylies, of Taunton, Massachusetts) arrived in 1846, and were large red or buff fowls without feathered legs. The 'Great Malay' illustrated in Dr Bennett's book were tall, but with more profuse plumage and higher tail carriage than exhibition Malays. The description and illustration of the 'Wild Indian' clearly suggests that it was an Asil hen.

His description of 'Plymouth Rocks' virtually admits that they were not yet a proper breed: 'Their legs are very large, and usually blue or green, but occasionally yellow or white, generally having five toes upon each foot. Some have their legs feathered, but this is not usual.' The rest of the description and the primitive illustration suggests that they were roughly the same shape as the later utility type Rocks, and Black-Red/Asiatic Partridge in colour. They died out by 1860.

Seabright Cochin (*N.B.* This was the spelling of Sebright used.)

Protoytpe Silver Laced Wyandotte were jointly produced by Mr A.S. Baker, Rev. Benson and Mr John P. Ray of New York. They started about 1864–7, during the closing stages and aftermath of the Civil War, initially with an oversized Sebright bantam cock, a Buff Cochin hen, then adding Silver Spangled Hamburghs, Dark Brahmas and Bredas. Some of the details were a cause of argument in later years, and most are described in the 1905 (American) edition of Harrison Weir's *Poultry Book*. They were a very variable lot of birds, this being early in the breeding process

Sherwood

Developed during the 1850s by the Timberlake family on their Virginia plantation called 'Sherwood', they survived the Civil War, and were promoted by the 'Burpee Company' (seed and agricultural merchants) in the 1880s and 1990s. Starting from crosses of White Georgia Games (an American Pit Game strain) and Asiatics (Light Brahmas and White Cochins?), Sherwoods resembled White Croad Langshans in every respect except for having yellow feet. They were believed to be extinct by 1920, but probably were at least a decade earlier.

White Wonder Fowl

A breeder (name not found) in the state of Maine developed this white-plumaged, round-bodied, slightly feather-legged breed that had a few years of local popularity in the 1890s. They were never recognized by the APA, and probably died out by 1900.

Winnibago

Joseph McKeen of Winnibago Farm, Wisconsin, crossed Rose-combed Brown Leghorns, Pea-combed Partridge Cochins and Buff crossbred (Gold Sebright × Buff Cochin) hens. These crossbreds were mated to American Seabrights (*see above*). After Silver Laced Wyandottes were standardized in 1882, Mr McKeen wisely changed the name of his breed to Gold Laced Wyandotte. Therefore, Winnibagoes are far from extinct, although the name has long gone.

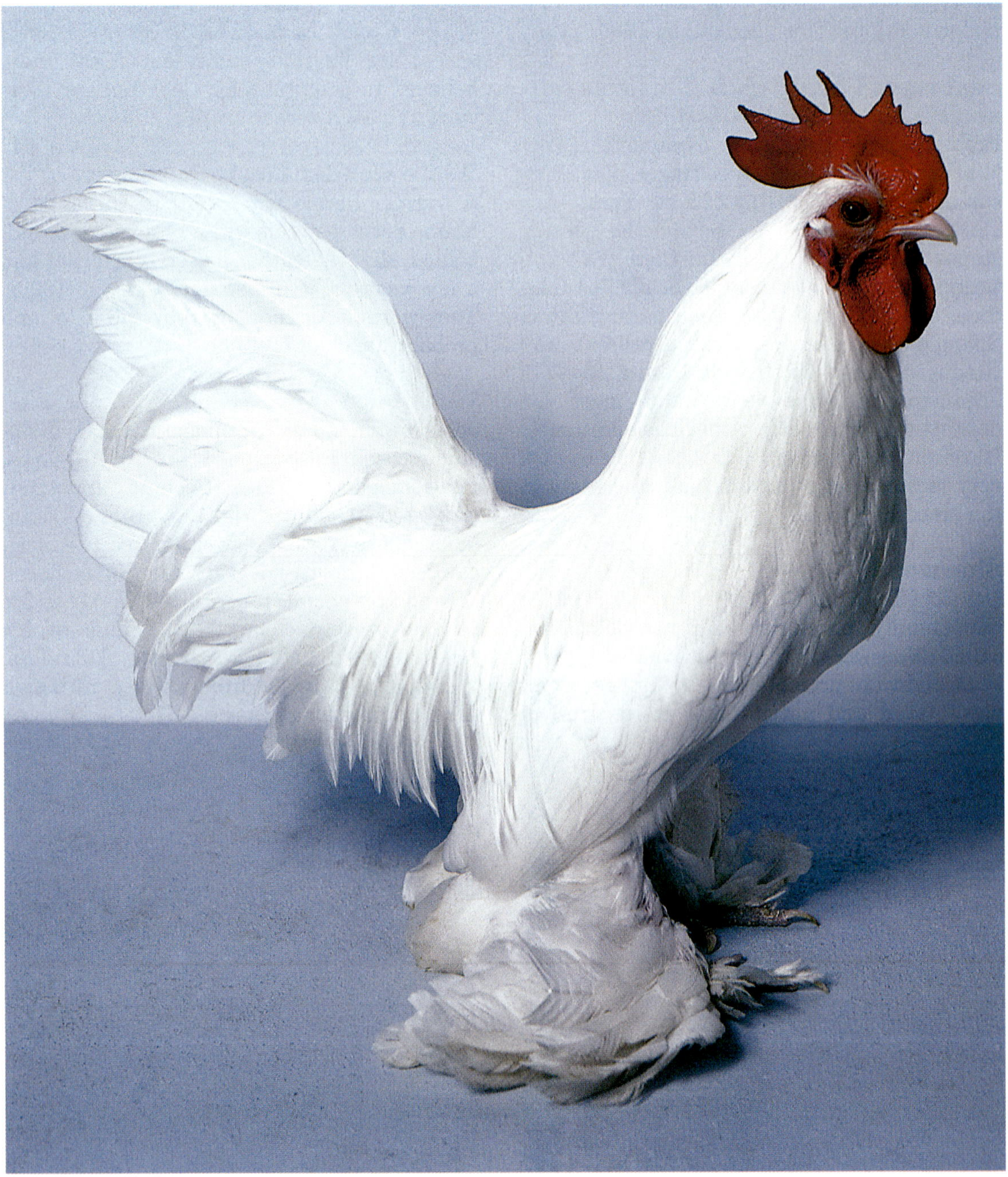

CHAPTER 13

True Bantams

True Bantams are defined in the UK as those breeds that are only bantams, in contrast to the miniature versions of large breeds. A lot of them are very popular in many countries around the world, too popular for inclusion in 'Rare Poultry Breeds'.

Popular True bantams not covered in this book include:

Barbu d'Anvers/Antwerpse Baardkriel, Barbu d'Uccle/Ukkelse Baardkriel
Barbu de Watermael/Watermaalse Baardkriel, Dutch/Hollandse Kriel, Japanese/Chabo
Pekin (only named as such and classed as a True Bantam in the UK. Elsewhere they are bred with a more upright carriage and classed as Cochin bantams)
Rosecomb/'Bantam' in Germany/'Java' in The Netherlands
Sebright.

True Bantams included in this chapter are:

Belgische Kriel (Belgian Bantam) – Belgium
Booted (UK)/Federfüßige Zwerg (Germany)/Sabelpoot (Netherlands)
Burmese – Burma and the UK
Deutsches Zwerghuhn – Germany
Gele van den Mehaigne – Belgium
Nankin – UK from unknown Asian imports
Ohiki – Japan
Pictave – France
Swedish Bantam – Sweden
Toumais/Doomikse Kriel – Belgium.

See Rumpless Breeds chapter for: Barbu d'Everberg/Everbergse Baard Kriel, Barbu du Grubbe/Grubbse Baard Kriel, Ruhlaer Zwerg-Kaulhuhn/Ruhlaer Rumpless, Uzara-O.

BELGISCHE KRIEL (BELGIAN BANTAM)

These are conventional Black-Red/Partridge bantams, and can be regarded as the Belgian equivalent of Dutch/Hollandse Kriel, Pictaves, Swedish Bantams and others. Belgian fanciers started to select for a specific size and type about 1909, but they were not recognized and standardized until 1934.

Official weights are: cock 650g (23oz), cockerel 550g (19oz), Hen 550g (19oz), pullet 500g (17½oz). Each of these is 100g (3½oz) above the equivalent for Dutch Bantams. The few specimens that the author has seen seemed more than 100g (3½oz) heavier than Dutch. They have smallish, upright, single combs, red lobes and orange-red eyes. There is only one recognised colour – Black-Red/Partridge.

BOOTED/FEDERFÜßIGE ZWERG/SABELPOOT

Chickens with feathered feet were described by Ulisse Aldrovandi of the University of Bologna, Italy, back in 1600. There have been other reports of broadly similar birds all over

True Bantams

Belgian Bantams / Belgische Kriel, male. Photo: Hans Schippers

Europe in the centuries between then and the beginning of organized poultry showing in the mid nineteenth century. Before the shows, and formal breed standards, birds were selected for breeding, if they were selected at all, according to the personal preferences of each flock owner. As the alternative names of the breed given above suggest, the three main countries in which random feather-legged chickens were transformed into a standard breed were Great Britain, Germany and The Netherlands. British breeders were soon 'out of the game' however, and it was the many keen German and Dutch breeders who really gave us the breed we see today. They remain very popular in these two countries, although rare in the UK.

Before they became a uniform breed, there was lot of variation in the shapes and sizes of feather-footed bantams, some of which were recorded in *Wright's Book of Poultry*.

Wright quotes from Mr Hewitt, also well known as one of the pioneer Sebright breeders, who described the Booteds he kept (in the Birmingham area) back in the 1830s. These were about the size of Silkies, then halfway between large fowl and bantams (approx. 1kg (2lb) pullets, up to 2kg (4lb+) adult cocks). They were bearded and had large rose combs. Plumage colour 'closely resembled the Red Speckled Dorking'. This may have been either as present-day Speckled Sussex, or as Spangled Old English Game, or more likely, both. 'They were long in the wing and tail-feathers; the wings drooped nearly to the ground; but the greatest novelty about the bird consisted in the extreme length of the hocks and leg feathers.' 'The leg feathers of both cocks and hens usually measured six inches (15cm) in length, and once knew a cock that always, when in full plumage, measured seven and a half inches (19cm) in the leg-feather.' 'Their flesh and skin was as white as the Dorking's.' This strain was believed to have died out by 1872, when Wright's *Book of Poultry* was first published.

The same book also describes more conventional (small) White Booted Bantams, and mentions that Blacks had been known, but were then even rarer than the Whites. A

White Booted Bantams. Artist: J.W. Ludlow. Originally a free gift with Poultry *magazine, circa 1912–14.*

Mr Beldon had some Bearded White Booteds in 1871, which he had imported from Germany. All feather legged bantams were at a disadvantage when the shows started because they usually had to compete with the easier to prepare clean legged breeds in 'A.V. Bantam (Not Game)' classes.

William Flamank Erntwisle (*Bantams*, 1894) gave more details of the very varied range of 'Booted Bantams' in late Victorian England. He noted that there were both long and short-legged types at the shows. A (black and white) engraving shows a pair of irregularly mottled birds with very long foot feathering on very short legs, full body feathering and long upright tails, suggesting they might have been Pekin × Japanese.

Entwisle said the main varieties of Booteds were: 'Black, White, White-whiskered [White with beard], Speckled Black and White, Speckled Red, Dutch-bearded Booted.' He went on to name other feather-legged breeds. Speckled reds were like 'Black-reds that have been out in a snowstorm'. There are also tricoloured Booted, which like the Bearded, are chiefly of Dutch or Belgian origin – many of them marked very much like the Almond Tumbler pigeon.

There never was a British Booted Bantam Club, the breed coming under the general 'Variety Bantam Club' for some time in the early twentieth century. However, when Belgian Barbu d'Uccles were introduced to the British Poultry Fancy shortly before the First World War, a British Belgian Bantam Club was soon formed to promote both Barbu d'Anvers and Barbu d'Uccles. At the same time (circa 1900–14) Brahma and Cochin/Pekin Bantam classes were attracting entries from many keen exhibitors. Here are some

True Bantams

entry figures at some of the main shows of the period. Most shows would not have any Booteds entered at all.

Variety Bantam Club Show at Sheffield, December 1910. This comprised 453 entries in 68 classes. There were two classes for Booteds, six of each sex being entered. Four of the six prize cards (1st and 3rd M, 2nd and 3rd F) went to Mr Kenneth Ward from Yorkshire who was the leading Booted breeder in the UK for about thirty years (circa 1898–1928). At the same event there were twenty-six Brahma Bantams (four classes) and seventy-two Pekins (thirteen classes).

The Variety Bantam Club's 1911 event did not have classes for Booteds, but Mr Ward won both male and female AOV classes. In 1912, the Variety Bantam Club's Show, this time at York (handy for Mr Ward) had ten male and sixteen female Booteds. This time Mr Ward had to make do with second in both classes, the two firsts and one third prize going to O.S. and E.B. Binns.

All the main shows in 1913 had combined 'Booted or Belgian' classes. Mr Ward won several prizes, but as he was among the first to take up both Barbu d'Anvers and Barbu d'Uccles, we can only guess how many Booteds were still being shown. The 1914 *Feathered World Yearbook* includes photographs of a male and female White Booteds, both Mr Ward's. They appear to be much the same as today's birds. During the 1920s most breeders seem to have crossed Booteds with Barbu d'Uccles; even Mr Ward who did not have Booteds by 1930. Bantam expert, Harry Inman, writing in the 1930 *Feathered World Yearbook*, said the only Booteds he knew of in Britain were owned by Mr Walker of Ipswich, Suffolk.

Hardly any Booteds were seen at British shows from 1945 until about 1985, when stock and eggs from Germany and The Netherlands started to appear. Andrew Sheppy (Rare Poultry Society founder) obtained some about 1970, believed to be from an old English strain.

Booteds do not seem to have been any more popular in the USA than they were in Britain by the end of the nineteenth century. As in Britain, there were certainly plenty of feather-legged fowls about, some Silkie size, others much smaller, in the mid-nineteenth century, but as the shows became established Booteds were neglected in favour of Brahma and Cochin Bantams (not Pekins in USA) by serious exhibition breeders. T.F. McGrew published *The Bantam Fowl* in 1899 that gives a detailed account of the Bantam breeds of the period. His description of Booteds is a retelling (duly acknowledged) of Entwisle's 'Bantams' section on the breed, as are two of the illustrations. There are two more engravings of Booteds, one being by English artist J.W. Ludlow, and one (artist's name not clear) American. The pair of Booteds in this picture, a White cock and a Black hen, appear quite large and tall. If these birds were alive today, they would be considered more Croad Langshan Bantam than Booted Bantam.

Mr McGrew gave a few more details in 1905 (*The Poultry Book*, Revised US edition of Weir) on Booteds as bred in America. He confirmed that: 'The early day Booted bantam was of diverse kinds.' Whites were said to be the most popular variety in America at this period. McGrew also noted that although official (US) standard weights were similar to Sebrights and Rosecombs, in reality the Booteds were much larger.

Most other American poultry and bantam books seem to give near-identical descriptions. One might wonder how many of the authors had ever seen a Booted? Fred P. Jeffrey (in *Bantam Breeding and Genetics*, 1977) was better, he quoted from two real contemporary Booted breeders, Henderson (1959) and Kellogg (1968). American Bantam Association Yearbooks include a few breeders' adverts, mostly of the Millefleur colour variety. The author does not know the origin of these birds. An old US strain? German or Dutch? Modified in the USA from Barbu d'Uccles?

The breed was officially named as 'Nederlandse Sabelpootkriel' in the Netherlands in 1902 or 1903, initially just in the 'Porselein'

Porcelaine Booted Bantams. Artist: Kurt Zander

(Millefleurs) variety. More colours were added over the years, including 'Blauw Porselein' (= UK 'Porcelaine'); sixteen colour varieties by 1949. The 1985 Nederlandse Sabelpoot Club Show had 207 birds entered in twelve colours. Since many Booteds in the UK in recent years have come from The Netherlands, many British fanciers have started to use the name Sabelpoot.

Development of the breed was recorded in some detail in Germany when *Federfüßige Zwerghüßner, Die Zwerghuhn-Rasse der Welt* was published in 1987. Unusually for the time, this book was jointly published by the then separate West and East (DDR) German Federfüssiger breed clubs. The text suggests that there had been long-term links between the two breeders' groups. Perhaps the DDR Federfüßiger Club members had some friends in high places to enable them to bypass the infamous Berlin Wall?

The First German National Poultry Show at Düsseldorf in 1886 included three pairs of Millefleur Booteds. By 1896, breeding had progressed to enable an entry of seventy-four Booteds at the National Zwerghuhnshau (Zwerg = dwarf, the general term for Bantams in Germany), by then including Blackmottles and Whites as well as Millefleurs.

In January 1911, Paul Hahn wrote in two poultry magazines to propose the formation of a breed club for Porzellanfarbiger Federfüßiger Zwerghühner (Millefleurs Booted Bantams). The breeders who responded to this call met on 9 April 1911. They were all in favour of forming a club for Federfüßige Zwerghühners, but insisted that the new club cover all colour varieties, not just Porzellanfarbiger (Millefleurs). An official inaugural meeting was held on 19 November 1911, at which Paul Hahn was elected chairman and the breed standard that had been

True Bantams

Buff Mottled Booted Bantam, male. Photo: John Tarren

Buff Mottled Booted Bantam, female. Photo: John Tarren

provisionally published in August, was formally adopted. The first standard described the breed type (shape) and three colour varieties: Porzellanfarbige (Millefleur), Blauporzellanfarbige (= UK Porcelaine) and Chamoisfarbig (Buff with white mottling). The other colours were added gradually over the years, with progress obviously being held back during the wars.

Progress has been steady since 1962, when the breed club show had 325 birds entered. Entries ranged from 403 to 568 during the 1960s and 1970s; with the 1,000 bird barrier being broken in 1986. In addition to the ever widening range of colours, there were also bearded Booted kept by German fanciers instead of Barbu d'Uccles.

German breeders in the late nineteenth and early twentieth centuries referred to Blacks and Whites as 'English Garden Bantams'. These two colour varieties indeed seem to be British in origin, even Blacks and Whites in Germany and The Netherlands nowadays are smaller and more compact in build than the patterned varieties. The full range of varieties we see today are Dutch, German (and to an extent, Belgian) in origin. Some details of their development in Germany are given below, but unfortunately similar details of Dutch strains were not available.

Black Mottled Booteds, once called 'Houdan farbige', were bred by Gustav Hartung, from Grimma and Hugo du Roi, of Braunschweig (a.k.a. Brunswick). The latter breeder is better known as the creator of Red Saddled White Yokohamas, a variety that never existed in Japan. He died in 1910. By this time the white spotting on the black background was excellent, but type and size needed improvement. This was achieved by others in the following years by crossing with Black and White Booteds. In the process, some appeared with beards. Blue Mottleds were developed, along with self Blues, by E. Klein of Benin during the 1930s.

Buffs were made by Otto Müller, also in the 1930s. His strain died out during the Second World War, but had not been perfected anyway. A new strain made in the 1960s has not attracted many breeders. If the winning birds are a guide, even the best have colour faults such as dark shading in tail feathers.

Columbian Booteds were in the 1930 edition of the German Standards, but only a few were said to exist then, and these were not very good. This stock died out during the Second World War. The variety was made again during the 1960s – a mottled-breasted Silver Duckwing (Silberhalsig) cock mated to two Buff hens being the foundation. Some excellent specimens were awarded the highest grades during the 1969/70 winter German show season. Before then they had problems with creamy ground colour and weak neck striping.

White Booteds had obviously been bred in Germany since the mid-nineteenth century, and were usually credited as being of English stock. Although of excellent quality, the White variety does not seem to have had many exhibitor breeders, the various colour versions of the Mottled and Millefleur patterns attracting the most interest.

Barred Booteds were produced by Emil Leipner and Ernst Mensinger in the 1930s from crosses between Black Booted and Barred Cochin Bantams. A new strain of Black Mottleds appeared as a by-product. They have sometimes been called 'Cuckoo' rather than 'Barred'. The barring clarity is somewhere between that of Barred Wyandottes and Scots Greys.

Black-Red/Partridge Booteds were known in Germany in the 1920s and 1930s, but never in great numbers, and these had died out by 1945. They were remade in the 1950s by Phillip Müller and Hans Otto Neumann from (Millefleur?) Booteds and Black-Red/Partridge Deutsches Zwerghuhner (German Bantams). It is believed they were first exhibited at the 1961 German National Show.

Silver Partridge (Silver Duckwing) Booteds had a similar history, made in the 1930s, and again in the 1960s.

Birchens are a new variety, first recognized in 1969, and were made from Black Booteds, Birchen Japanese (Chabos) and Birchen Cochin Bantams.

Description

The wording of American, British, Dutch and German standards all differ slightly from

White Booted Bantam, male. Photo: John Tarren

White Booted Bantam, female. Photo: John Tarren

each other, and sometimes all seem to differ from some of the birds winning at the shows. Everyone agrees on the colours of plumage, eyes, shanks, lobes and so on, the smallish straight single combs and the well spread, fairly high tails that on males have 'sabre-shaped' sickles. It is on matters of size and shape that differences can be seen. Young birds usually seem taller and slimmer than older specimens. Black and White Booteds are (on average) slightly more compact than the rangier Millefleur and Mottled varieties.

Some people might wonder what, if any, difference there is between Barbu d'Uccles and Bearded Booteds? In many cases, the real answer is probably, 'not much'. Ideally a Barbu d'Uccles should be more compact than most Booteds (but very similar to most Blacks and Whites), and have thicker neck feathering with a well-developed 'boule', or swept-back section halfway down the side and back of the neck. Booted neck feathering is less profuse, which no doubt gives an impression of a taller, slimmer bird generally.

Booted Bantams must, practically by definition, have very long and profuse foot feathering all down the shanks and growing from all the toes. They also have long 'vulture hocks', stiff feathers projecting from the hock joints. To be exhibited successfully all the specialist housing and management techniques in relation to keeping feather-footed breeds must be employed. The foot feathering on Booteds and Barbu d'Uccles is more brittle, and therefore more easily broken, than that of Cochins/Pekins. Those intending to successfully show White Booteds that must, of course, be kept pure snow-white, should be realistic about the number of other varieties they will have the time to prepare in the week before a major show. Their 'other breeds' should be 'low maintenance' types, darker coloured and without feathered feet, long tails or crests.

Booted Colour Varieties

Millefleurs (in UK, Belgium and France) = Porzellanfarbig in Germany, Porselein in Netherlands.
The general ground colour is rich orange-red. Most feathers have a large round black spangle towards the end, terminating in a triangular white tip; the tail and foot feathering is black with white tipping. The white tips get bigger with each moult, so under-spotted young birds are often better in their second year, and some prize-winning youngsters become too 'gay' later in life.

Porcelaine (UK) / Blauporzellanfarbig (Germany) / Blauporselein (Netherlands)
The German and Dutch names for this variety are misnomers because this variety is caused by the Lavender gene (not the Blue gene) diluting the Millefleur pattern. The orange-red ground colour is diluted to delicate creamy ochre, and the black is replaced by lavender.

Blue Millefleurs
The Complete Encyclopedia of Chickens (Esther Verhof and Aad Rijs, 2003, The Netherlands) includes a photo of a (genetically) Blue Millefleurs hen. The 'blue' parts are not very different in shade from some darker specimens of Porcelaines, but the ground colour is an orangy shade, similar to that of normal Millefleurs. These are probably a new variety.

Citron Millefleurs
These have normal black and white spotting with a yellow ground colour. They have been known at least since 1915, but have only become a popular colour since the 1980s.

Silver Millefleurs
At first glance these look like Silver Spangleds (as Hamburghs), but on closer inspection the triangular white feather tips will be seen.

Black Mottled, Lavender Mottled, Buff Mottled
Ideally all these three varieties should have neat white tipping as that of Anconas. They

are never quite this tidy. Buff Mottleds generally have clear white tail and foot feathering, not buff with white tipping as might be expected.

The remaining colour varieties are all similar to their equivalents in many other breeds: Barred, Birchen, Black, Black-Red/Partridge, Blue, Buff, Lavender, Red, White, Silver Partridge (Silver Duckwing), Red-Shouldered Silver Partridge (Gold Duckwing).

BURMESE BANTAM

A few British fanciers bred and exhibited Burmese Bantams circa 1880 to 1900. Burmese then virtually vanished for almost a century. Andrew Shoppy (UK Rare Poultry Society founder) obtained some from an elderly fancier in Wiltshire about 1970 that were believed to have been the last survivors of the original stock detailed below. Andrew still had some of these at the time of writing in 2005, but Burmese are not very fertile, and Andrew has always kept a large collection of breeds. Some breeders in The Netherlands made a 'facsimile' strain during the 1980s.

The only significant account of Burmese in the Victorian era found by the author is in W.F. Entwisle's *Bantams* (1894), and the only pictures, a drawing in Harrison Weir's *Our Poultry*, and an engraving by J.W. Ludlow in Edward Brown's *Poultry Breeding and Production* (1929) that shows Burmese, but is mistakenly captioned as 'Sultans'.

They were first sent by a British army officer in Burma to another officer in Scotland (possibly both in The Black Watch regiment) sometime in the 1880s. These had pure white plumage. The hens had died before Entwisle got hold of them, the Scottish climate obviously being a shock to the birds after the heat of Burma. This cock had a large, feathered

White Burmese Bantams, pair. Artist: J.W. Ludlow. From Poultry Breeding and Production *by Edward Brown, where it was incorrectly captioned 'Sultan Fowls'.*

True Bantams

crest and a small single comb on his head. He had low wings, a long tail and very short, heavily feathered legs. Apart from the feathered legs, all these other points suggest a relationship with Japanese (Chabo) Bantams. His foot feathering was said to be very long, quite the equal of the best Booted/Sabelpoot/Federfüßige Bantams today. Their skin and shanks were yellow, another Chabo characteristic. As Entwisle only had a cock, he obviously needed some 'near enough' hens to put with him. Luckily, he had already made Sultan Bantams (using undersized large Sultans, White Booted Bantams and White Polish Bantams). The resulting chicks came in a range of colours apart from white. In the breed standard given in Entwisle's book, the colour varieties given are: White, Black, Brown, Grey and Speckled. No further details of the last three varieties are given. The weights given are similar to that of Japanese (Chabos).

The new Dutch strains include the main two colours, Blacks and Whites, along with Buffs, Black-Tailed Buffs and Brown Reds. From the few photos seen, they still have two key faults: horned combs instead of single combs, and insufficient length of tail and foot feathering.

DEUTSCHES ZWERGHUHN (GERMAN BANTAM)

Willtheim Müller, of Magdeburg (died 13/1/1956) produced the Wildfarbig Deutsches Zwerghuhn in 1917. German fanciers had made a major contribution to the development of a lot of breeds, but their efforts may have gone unnoticed outside the world of poultry enthusiasts. Perhaps feelings of German nationalism created a need for a clearly German ornamental bantam breed, and Herr Müller's creation was the first attractive enough to attract widespread interest. They remain a very popular breed in Germany (and Austria?), but are little known in other countries, although a few arrived in the UK about 2000.

When they see a German Bantam for the first time, most poultry fanciers would say that they look like a half (single-combed) Yokohama (= Phönix in Germany) crossbred, which is pretty much what they were. The other breeds used included Rosecombs, OEG and some local crossbred bantams that were believed to be similar to Belgische Kriel or Pictaves.

Description
These are roughly midway in size between Dutch and Yokohama (= German Phönix) Bantams, with males at about 750g (26oz) and females weighing 600g (21oz). Their shape and style is also halfway between these two breeds. Tails are longer and held at a lower angle than on Dutch, but not as extravagant, or as low as Yokohamas. In this respect they would be a much easier to manage variety for the many British (and other) people who have Yokohamas as ornamental garden pets, but have neither the time nor inclination for the special management needed to keep very long tail and saddle feathers in perfect condition. Being larger than the Dutch, German Bantams will be hardier and lay eggs of a more useful size. They have small, upright, single combs, white ear lobes and slate-blue shanks and feet.

'Wild Colour' (Wildfarbig)
The original variety made by Herr Müller is a dark Black-Red. Males have very glossy plumage in greenish-black and rich mahogany. Females are a dark variant of the normal partridge pattern. The neck hackles might be described as black with a gold edging instead of the normal gold with a black centre stripe. Back and wing plumage has the normal brown with black peppering, but there are also black crescent spangles, almost (but not quite) lacing. Main tail feathers are black. Breast plumage is a shade darker and more brownish than the salmon seen on other Partridge females, and there are also some black crescent-shaped spangles. Red Dorking females should have similar markings, but seldom do. This plumage pattern is otherwise

mainly seen on some Shamo and related breeds.

Several more colour varieties were developed during the 1960s and 1970s, all of which are similar to their equivalents in Dutch Bantams: Birchen, Black, Black-Mottled, Black-Red/Partridge ('Goldhalsig'), Blue-Red ('Blau-Goldhalsig'), Blue-Gold Duckwing ('Blau-Orangehalsig'), Blue-Silver Duckwing ('Blau-Silberhalsig'), Buff Columbian, Columbian, Gold Duckwing ('Orangehalsig'), Lavender, Pile ('Rotgesattett'), Millefleurs ('Porzellanfarbig'), Silver Duckwing ('Silberhalsig'), White.

GELE VAN DEN MEHAIGNE

This breed was developed by M. Herregodts about 1940. The Second World War was obviously not the best situation for a new breed of bantam to take hold. They were fully described in Willem and Brandt's 1971 book, but not in many other places. Wandelt and Wolters (1998) also covered them, pointing out that there were none at the 1989 Europaschau at Ghent, Belgium. At this special event, one would have expected every Belgian breed to be represented. The author does not know if any exist now. Any revival is likely to be in their home area around the towns of Novilles/Mehaigne and Eghezée, south-east of Brussels.

Gele van den Mehaigne are (or 'were', as the case may be) a simple black-tailed-buff light breed with a single comb, white lobes and conventional light-breed type.

NANKIN

The Rev. W. Wingfield and G.W. Johnson, in *The Poultry Book* (1853), wrote:

> The Nankin or Yellow Bantams, appear to have been among the earliest importations; and, though now seldom considered as show birds, their clear plumage and active figure still procure for them many admirers. Their prevailing colour is the pale orange-yellow of the Nankeen (a corruption from Nankin), a cotton material formerly much in vogue for summer wear in this country, as also in India. The hen has usually some dark markings on the hackle (but the less the better), and the tail is often tipped with black; while the male bird shows an intermixture of red and dark chestnut on the back and wings, the tail being black and well arched. Both sexes have short dark legs, and a double [rose] comb.

This passage concludes by saying that many Nankins had single combs, and that their owners considered these as equally correct.

Even then, when poultry shows had just started, Nankins were already being neglected. They were regarded as 'the common sort', and most fanciers wanted something exotic.

Lewis Wright went into more detail in the earlier editons (1872–91) of his *Poultry Book*: 'Mr. Oliver E. Cresswell brought this Bantam again into notice by a beautiful pen exhibited at the Crystal Palace "Bantam" Show of 1872, and has at present a fine yard of them.'

Wright includes a passage from Mr Cresswell, of which the following indicates the position of Nankins in the poultry keeping world at the time:

> This lovely breed of Bantams has been sorely neglected of late, but I do not think any catalogue of Bantams complete without mention of them. There are those, though not among the class of exhibitors, who, knowing their beauty and merits, have kept and bred them carefully through a long series of years; and excellent specimens may sometimes be seen at gamekeepers' cottages in remote districts, where they have long been kept as foster-parents to partridges.

Thus Nankins were kept to perform the same duties as Pictaves were (and still are) in France (*see* p.247).

Mr Cresswell went on to describe Nankins. Once again both single and rose combs were known, Mr Cresswell preferred the rose, but he said 'that the best Nankins of thirty years

True Bantams

ago had single combs'. Thirty years ago would then have been circa 1840. Shanks and feet were then, as now, either blue or white. Mr Cresswell favoured blue, but most of his stock had white legs. At the shows the two comb types and leg colours did not matter, as long as all the birds matched in trios classes.

In Wright's updated (1902) *New Book of Poultry* he wrote that they were virtually extinct. W.F. Entwisle (1892) also said that they were almost extinct, Mr Cresswell and Mrs Ricketts being the only significant breeders. All the main poultry and bantam books, including the *Feathered World Yearbooks* (1910 to 1938), thereafter speak of Nankins in the past tense, if they mention them at all. Nankins were said to be historically important, being used to make Gold Sebrights and several buff, or partly buff, breeds. They were said to have been roughly the same size and size and shape as Rosecombs. The author has only found one photo of Nankins in 'the old days', this being in *The Encyclopaedia of Poultry*, J. Stephen Hicks (circa 1921, vol.1, p. 42). These were described as 'the property of Mrs Cross', and show birds that are small, with tight body feathering and large well-fanned tails. Generally they look more like American-type Old English Game Bantams than most of today's (2005) Nankins as shown in the UK.

The idea that there are flocks of rare and ancient breeds of poultry (and other livestock) unknown to the 'establishment' running around in remote smallholdings is generally a modern myth; but in this case it really happened, well nearly. They were 'found' by Mrs M.F. Peters of Ringmer, near Lewes, Sussex; and had been bred, quietly, by Mr Fred Martin of Wisbech, Cambridgeshire, in 1955. This was not quite the 'chance encounter in a country lane' situation that the romantically inclined might dream of. Both Mrs Peters and Mr Martin were members of the old Houdan and Creve-Coeur Club, and it was in this club's September 1955 newsletter that the revival of Nankins was reported. This strain of Nankins had been maintained for a long (unspecified) time by Mr Martin and several of his relations. This suggests that they were spread over a number of sites, which presumably accounts for their reported excellent fertility.

The next significant article on Nankins, by Andrew Shoppy (RPS founder) appeared in the 1971 *PCGB Yearbook*. In it Andrew said that he obtained his stock from Mrs Peters, who by then had built her flock up to about

Nankin Bantam, male. Photo: John Tarren

Nankin Bantam, female. Photo: John Tarren

sixty birds. However, by the 1970s Nankins had increased in size and had lost the strutting carriage of the Nankins in the 1921 photo.

Present-day Nankins are correct as far as plumage colour is concerned, and the two comb types and shank/foot colours have always existed side by side. Size and carriage remain a problem, and enough birds should be bred each season to allow for selection.

OHIKI

This curious breed from Japan has only become known to fanciers around the rest of the world since the 1990s. At the time of writing (2005) there were a few Ohiki in England, notably exhibited by Richard Billson (RPS Secretary), and no doubt there are others elsewhere in Europe.

Despite only being known to the outside world recently, Ohikis are believed to have existed in Japan since the mid-nineteenth century. They were developed in the Prefecture Kochi on the southern Japanese island of Shikoku. It is assumed that they were made from some long-tailed fowls (O-Naga-Dori, Minohiki and the long-tailed long-crower Tôtenkô) and small bantams such as Chabo or some Pekin-like birds from China.

Older Ohiki cocks have very long saddle and tail feathers that drag along the ground, which is perhaps a contrast with the generally rounded and compact form of females and young males that have yet to grow their long tail feathers. The hens have a broad and well-developed cushion rising to the tail that is longer than that of (say) Pekins or Wyandottes, but not as long as might be expected considering the tails of the adult cocks. Their legs are fairly short, appearing to be very short on profusely feathered older birds. Ohikis have white ear lobes and small to medium-sized single combs. Three colour varieties are recognized, White, Silver Duckwing and Black-Red. The markings of the Duckwing and Black-Red females are distinc-

Mr Yamashita and his Ohiki in Japan. Photo: Julia Keeling

tive, and may be the result of the eb gene. Silver females have a creamy-white ground colour, black centre striping in neck hackles and fine grey peppering on their back and wings. The ground colour of the Black-Red hens' plumage is rich yellow in the neck, and light wheaten brown elsewhere. Weights range from 600g (21oz) pullets to (traditionally, in Japan) 937g (33oz) adult cocks.

There are probably a great many 'Special Management Requirements' to produce a really long-tailed Ohiki cock, but the author does not know what they are.

True Bantams

Tournais/Doornikse Kriel, male. Photo: Hans Schippers

PICTAVE

Pictaves originated from an area of western France, south of the Loire river. Roughly from St Nazaire down to La Rochelle on the coast, and inland from Angers in the north-east and Poitiers in the south-east. They are a very natural looking, single-combed, Black-Red/partridge plumaged breed, so roughly similar looking birds may have been running around French farms for many centuries. Pictaves were bred along more systematic lines to produce this specific breed from about 1900. Female plumage colour has been fixed as roughly partridge marked all over, with no salmon breast. The original utility function of Pictaves was as broodies for hatching partridge and pheasant eggs for shoots. Pullets can be as small as 500g (17½oz), for those partridge eggs, with old hens up to 800g (28oz) for the pheasant egg hatching. Cockerels are 700g (25oz), cocks 900g (32oz) Pictaves have red ear lobes and greyish-white shanks/feet.

SWEDISH BANTAM

This is another localized Black-Red/Partridge, natural-looking bantam, which like the French Pictave was mainly kept for incubating partridge, pheasant or any other eggs required. Also like Pictaves, very similar-looking birds could have existed for a very long time before they were given a name and bred to uniform type and colour. This process is believed to have started in the 1920s, but a breed club was not formed until 1953.

Male Swedish Bantams look like slightly under-sized Welsummer Bantams except for having solid black breasts, and a little more neck striping. Their small single combs are typically of the 'flyaway' form. The females are perhaps more noticeably smaller, more 'gamey' and less of a layer type than Welsummer Bantam hens, perhaps just a shade lighter in colour.

TOURNAIS/DOORNIKSE KRIEL

Toumais/Doornik is a town in southern Belgium, just over the border and to the east of the major northern French city of Lille. They are normal-shaped, single-combed bantam of Spangled (white spots over Black-Red/Partridge) plumage pattern. As with the Belgische Kriel, Pictaves and Swedish Bantams described above, they were bred for a long time as broodies to no particular type or colour before they became a standard breed. In the early days they were usually larger (1,000g (35oz) pullet to 1,500g (53oz) adult cock) than the current standard form (650g (23oz) F and 750g (26oz) M). They were first described as a breed in the fanciers' paper *Union Avicole* issue of 21 September 1913. Since then, they seem to have remained rare – kept going by a few enthusiasts (for example, Nadine Tassheff and Jos Bruynooghe in the 1980s). Whenever inbreeding has been a problem, Spangled Old English Game Bantams have been the traditional cross used. Toumais/Doornikse have red ear lobes and white skin, shanks and feet.

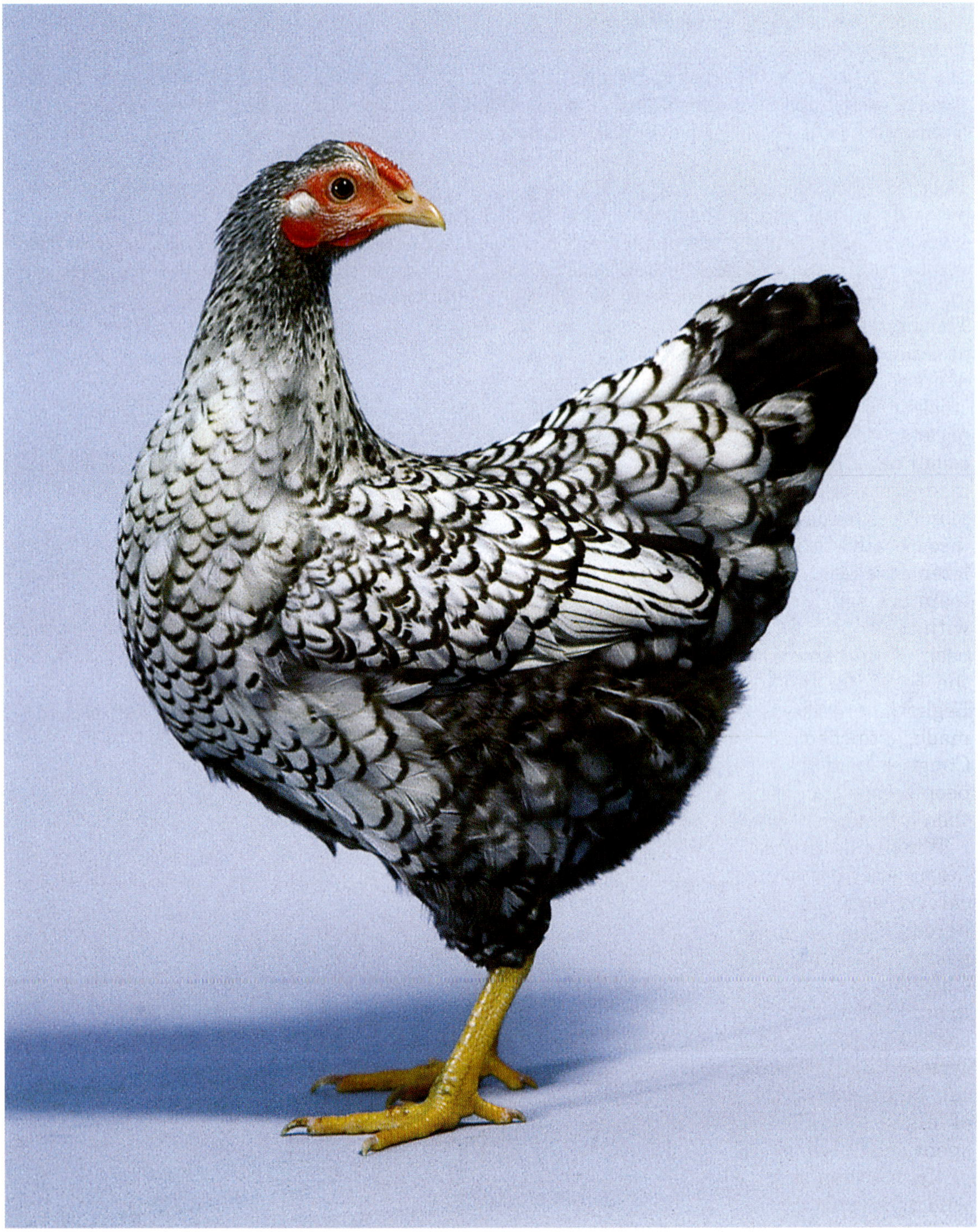

CHAPTER 14

Autosexing Breeds

The first Autosexing Breeds were developed at Cambridge University by some of the pioneers of the science of genetics. Day-old chicks of the Autosexing Breeds can be sexed because the cockerel chicks are a lighter colour than their sisters. The scientists who made them thought the principle would be of major economic importance, but it was not incorporated in the hybrid broilers or layers later developed in America. The Cambridge team seemed not to have been as in touch with the poultry industry as they might have been. Autosexing Breeds were named after the breed they were based on, Legbars on Leghorns and so on. Cambars were the first made, launched at the 1930 World Poultry Congress held at Crystal Palace, London. The people involved were world leaders in the then infant science of genetics.

William Bateson (1861–1926) studied wildlife around the world in the earlier part of his career, with an emphasis on variation and heredity. He returned to Cambridge in 1898, where he started working with poultry. Chickens are favourite animals for scientists to study because they are fairly cheap to buy, easy to keep and breed, have a fairly short life cycle (with artificial lighting, two generations can be bred in a year) and, most important of all, are economically important. Money spent on studying chickens is less likely to be criticized than money spent on studying a wild species from a remote country. Bateson was made Cambridge's first Professor of Genetics in 1908, indeed he invented the word 'genetics'. If it were not for William Bateson the study of inheritance may have been delayed, as he found and translated Gregor Mendel's studies of sweet peas from the original (neglected) German. Gregor Mendel (1822–84) was an Austrian monk and (part-time) naturalist, and therefore not part of the main scientific 'circuit'.

By 1904, Bateson had been joined by a young recruit, R.C. Punnett, who went on to become Head of Department after Bateson left in 1909 to become Director of the John Innes Horticultural Institution in Surrey. Between 1904 and 1909, Bateson and Punnett made some of the basic discoveries of poultry genetics, including the genetically different types of white plumage, the incompletely dominant blue gene (as Andalusians) and the genetic relationship between single, rose, pea and walnut combs.

The third key person at Cambridge's Agricultural Research Department was Michael Pease, who joined the group in 1922. Sex-linked crosses were a hot topic then, and the team put on a display of the Gold × Silver mating, usually Rhode Island Red × Light Sussex, at the 1922 London Dairy Show. Many readers will be surprised that this mating was 'new' in 1922, assuming that this mating was used long before then. According to the famous Dutch genetics expert Dr A.L. Hagedoorn, some Dutch and Belgian breeders were regularly using Gold × Silver matings at least as far back as 1905. Both groups were mating gold and silver varieties within a breed, Assendelftses in the Netherlands, Campines in Belgium. Both of these are light breeds, suitable only for egg production. Dr Hagedoom recalled that back in 1905–06 some unscrupulous Belgian breeders sent boxes of white chicks, all cockerels of course, to the Paris markets. It is not why gold × silver sex-linked mating did not become more widely known before 1914. Perhaps it was simply a case of people not realizing the potential of something they half knew already. Barred Plymouth Rocks sometimes produced black chicks, and these were always pullets; but no one knew why.

Following the Dairy Show, some British breeders asked the Cambridge team if it would be possible to create a single breed that would have different coloured male and female chicks. Punnett *et al.* had no idea how to do this. They eventually made the key discovery by accident. There were several reasons why the breeders wanted a single breed rather than producing sex-linked chicks by crossing:

- Poultry farming in the UK, even commercial egg and meat production was still tied up with the Shows in 1922. The separation between farming and showing had started, but most of the people involved, farmers and 'the experts' still tended to assume 'pure breeds' were superior to 'mere mongrels'.
- Show prize money and the prices charged for show birds were still high enough to be worth serious consideration.
- There were very few really large-scale poultry-breeding operations in Britain then, most flocks were no larger than those of hobbyists today. With sex-linked crossing, two pure breeds have to be kept as well as the sex-link breeding pens and rearing facilities for all stock not sold as day-old chicks. Both Rhode Island Reds and Light Sussex were popular varieties then, but serious breeders would prefer to have stock from proven strains rather than rely on being able to buy-in breeding stock of good enough quality each year. To keep all this breeding stock would have required many thousands of the final crossbred day-old chicks to be hatched for economic viability, and hardly anyone was operating on that scale then.

Before the Autosexing Breeds are described individually, there is one more person who deserves to be mentioned, John Croome. John was a student at Cambridge in the 1930s when these breeds were being made. He was still breeding some of them when he died in 1988. The Autosexing Breeds Association was formed in 1943, and John was on its committee from 1959, eventually becoming the Chairman. In 1972 the ABA was absorbed by the Rare Poultry Society (formed 1969), with John as RPS President from 1984–98.

The individual Autosexing Breeds are described in approximately chronological order. After the first few from Cambridge, other breeders added more, some from overseas.

CAMBAR

As was indicated above, scientists were only beginning to understand the genetics of poultry, or the genetics of anything for that matter, in the 1920s. Two of the plumage patterns being investigated then were 'Barring', the sex-linked type as seen on Barred Plymouth Rocks and 'Autosomal Barring', the pattern of Campines. In the course of experimental crossing between Barred Rocks and both Gold and Silver Campines, the team discovered the autosexing principle by accident.

On Barred Rocks, the barring gene is acting on birds that would have all black plumage if the gene was absent. There is some difference in colour between the sexes of day-old Barred Rock chicks, but the differences are not reliable enough for accurate sexing. The lightest chicks will be cockerels, and the darkest pullets, but there are a lot in between that are not sexable. When the barring gene is present on chicks that have one of the other genes on the 'E. locus' (that is, other than 'Extended Black'), the difference is more marked and more reliable. Experienced breeders of Crete Pit Game probably knew this a century before Punnett and Pease, but like the Belgian Campine and Dutch Assendelfte breeders, they failed to realize the potential importance of this knowledge.

The Cambridge team were well aware of the significance of this discovery, they had been looking for it since 1922. They had made enough progress by the time of the Third World Poultry Congress at Crystal Palace, 22–30 July 1930 to put on a display of adult 'Gold Cambars' and their chicks. 'Silver Cambars' were added to their displays at World Poultry Congresses at Rome in 1933 and Leipzig in 1936.

Although all the Autosexing Breeds included some Barred Plymouth Rock parentage, they were soon made to be virtually another colour variety of the other breed involved. Welbars are another colour of Welsummer in all but name. This was never true of Cambars, however; they never did look much like Campines. An adult Cambar cock weighed 8lb (3.6kg), compared to 6lb (2.7kg) for an ideal Campine cock, and very few Campine cocks actually weigh this much. Cambars were medium-sized dual-purpose birds, as one would expect from a mix of a light layer and a heavy table breed.

The Cambridge team applied to the PCGB for Cambars to be recognized and standardized in June 1946. This was done because breeds had always (since 1876 anyway) been standardzed, even those, like Cambars that were never intended to be show birds. Their first application was rejected because they were considered to be too much like the already standardized Golden Cuckoo and Silver Cuckoo Marans. There were meetings between Michael Pease and PC Council representatives that resulted in Cambars being recognized in 1947, with a promise from Cambridge that they would try to fix white ear lobes (the first Cambars had red lobes, from the Rocks) and a more Campine-like general shape.

Cambars failed to attract much interest, soon being supplanted by Legbars and Rhodebars. ABA *Autosexing Annuals* throughout the 1950s showed this in the breeders lists. By 1964 there were only two breeders, John Croome and R.G. Silson. It is not known when John Croome, the last breeder, finally gave up on them.

Description

Should anyone try to make Cambars again, which is probably not a good idea, they were very much a conventional medium-weight utility chicken in overall shape with upright single combs. Ear lobes should have been white, but it is not known if any really were. Eyes were 'rich bay' and beak, shanks and feet were white. The description of Gold and Silver Cambar plumage pattern is as vague and fuzzy as the plumage pattern was.

LEGBAR

Legbars, obviously based on Leghorns, were the second Autosexing Breed to be made, and during the lifetime of the ABA (1943–72) was the most successful. The Cambridge team had soon become aware that Cambars would never be commercial layers or table birds. Any commercial laying Autosexing breed would have to be Leghorn-based. Legbar development started in 1932, by which time the best laying strains were Whites, useless for Autosexing. Fortunately, there were still some utility strains of Brown Leghorns in the 1930s.

Gold Legbars were standardized by the PCGB in January 1945, with Silver Legbars following in December 1951. Work was able to continue through the Second World War because the Government thought it important in this era of 'Dig for Victory' and rationing.

A Brown Leghorn cock was mated to Barred Plymouth Rock hens, both from utility strains. Just two cockerels were retained for the next stage, and these were barred, possibly with a little gold in their neck and shoulders. They were mated with Brown Leghorn hens, and several hundred chicks were reared. This was necessary because, as expected, they came in a wide range of plumage colours and patterns.

All 'un-barred' birds were discarded from the programme, whether they were nearly all black or had some kind of pattern from their Brown Leghorn ancestors. Of the birds bred with barring, there was considerable variation. Those with some type of underlying Black-Red/Partridge were kept, and those that were barred over black (as Barred Rocks) were discarded. There were still a fair number of both sexes with acceptable plumage colour and pattern that allowed some scope for selection for Leghorn type, size and characteristic white ear lobes.

Males and females of this generation were mated together; and then gave:

Light-coloured males – the desired Gold Legbar male pattern, with two barring genes

Darker coloured males – discarded. These were very attractive, as the exhibition 'Crele' pattern on several breeds, but do not 'autosex'

Cuckoo Partridge (Crele) females – the desired Gold Legbar pattern

Un-barred Partridge females – similar to Brown Leghorns – discarded.

'Crele' is the traditional name, especially among Game Fowl breeders, for the Cuckoo-Partridge pattern. The darker-coloured males look very attractive, and are the favoured version at the shows on Crele Game, Crele Dutch, and so on. However, they have one barred and one non-barred gene, resulting in a mixture of chicks like the mating above. The darker Crele cockerel chicks are the same colour as Crele female chicks at day old. Once made, additional stocks of Gold Legbars could be quickly multiplied by mating Gold Legbar males with Brown Leghorn females. Back in the 1930s and 1940s there were plenty of Brown Leghorns available. They may not have been quite as popular as Rhode Island Reds or White Leghorns with poultry farmers, but Brown Leghorns were still a viable commercial laying variety. The result of this mating was 50 per cent dark Crele males, all discarded, and 50 per cent Crele, that is, Legbar, females.

Although the basic process as far as plumage colours are concerned is straightforward, it took a long time to establish Leghorn size, shape and utility characteristics. Mrs B. Sager of Southampton complained that many were still too much like Rocks in the 1957 *Autosexing Annual*, twelve years after Legbars had been standardized.

Silver Legbars were standardized in 1951. As Barred Plymouth Rocks are genetically silver, there may have been some silver chicks in from the early stages. The deliberately made Silver Legbars were started from a Gold Legbar × Silver Cambar mating. In order to improve productivity, crosses with utility White Leghorns were tried, and this would have given some Silver Legbars.

Both Gold and Silver Legbars were exported to several countries during the few years it was thought they had some commercial potential. They were sent to Australia, Kenya, New Zealand, South Africa and the USA, but made most impact in Germany. A cock and three hens were sent to an agricultural institute, the Landesgeflugelzuchtenstalt at Hohenheim near Stuttgart in 1938. They were still going post-war, probably being restocked from Cambridge (1948), and in 1951 a German Autosexing Breeders Association was formed, initially with thirty-eight members. They changed the name of the breed there from Legbar to Kennfarbig Italiener. (kennbar = recognizable, farbig = colour, Italiener = the German name for Leghorns, a result of them having first arrived in Germany overland from many parts of Italy. They were first imported to USA and UK from the port of Leghorn/Livorn.) Germany soon had several more Autosexing Breeds, some from the UK, others made in Germany. Gold Legbars still (2005) exist in the UK, but are rare. Silvers have not been seen for some time, but would be easy to make by crossing Gold Legbars with Silver Duckwing Leghorns.

Description

All details of the shape and size of Legbars were as for Leghorns. It must be pointed out here, as this book does not cover Leghorns, that the British Leghorn Standard described the slightly smaller utility type with well spread tails. The larger 'Exhibition Type', seen at its most extreme in some British strains of large White Leghorns, with very large combs and tightly whipped tails, were never officially described. In large British Leghorns, the Exchequer variety can reliably be said to have only ever existed in the utility type, so their shape can be taken as a fair ideal. Although both single and rose combs are allowed in Leghorns, only single combs were ever allowed in Legbars.

Gold females are very attractive, with salmon breasts and softly barred, grey and brown, on their back and wings. Neck hackle feathers are pale gold with black centre striping broken into barring. Gold males have soft cuckoo barred breast, legs and tail. They have light straw and gold barred (with a little black centre barring) neck and saddle hackles. Gold Legbar males look pretty enough, but their markings are too irregular to suit the tastes of most regular show exhibitors.

Similarly, Silver females are also much more attractive than their mates. They have salmon breasts like the Gold females, but Silvers' neck, back and wing plumage is an attractive mixture of black, grey and white, without the gold or brown shades of Gold Legbars. Likewise, Silver Legbar males have the same cuckoo 'undercarriage' and tail as the Golds, but top colour is mostly silvery white with some black centre striping broken into 'dashes' by the barring gene. They are less attractive than Gold males, which is a shame because it means we (judges) are unlikely to see any of the attractive females.

CREAM OR CRESTED LEGBAR

These have a distinctive plumage colour, a small head crest, and they lay blue-shelled eggs, a legacy from the Araucanas in their make-up. This is why they can be considered as a separate breed, rather than just another variety of Legbar. Blue eggs are a major attraction at present (2005), and although very few are seen at shows, they are becoming widely available and some pure breed poultry auctions have many lots of them, which attract brisk bidding.

Readers with a cynical view of the world will not be surprised to learn that this, the most popular of the Autosexing Breeds, was made by accident. In 1939, Michael Pease was trying to make an improved line of normal Gold Legbars by crossing some hens with an inbred, utility-type White Leghorn cock from the Cheshire College of Agriculture at Reaseheath. He was using the 'hybrid' method of crossing inbred lines to get improved productivity. White Leghorns have the Dominant White gene, which means they carry several other (unseen) plumage colour and markings. Two off-white pullets were kept from this mating that were mated with a Gold Legbar cock (from the same strain as the 1939 hen) in 1940 and 1941. A total of 118 chicks were hatched, of which sixty-nine were white, or nearly white, and forty-nine were coloured, more or less like normal Gold Legbars. Some brother-sister matings were tried from these forty-nine coloured birds to reveal any recessive genes carried by the White Leghorns from Reaseheath. One of the 1940-bred cockerels produced some unexpected creamy-coloured chickens. What Michael Pease found really interesting was that there was still a clear colour difference between male and female chicks, but when adult, although there were natural sexual differences in plumage pattern, the cockerels did not stand out as much lighter when the flock was viewed from a distance. These first Cream Legbars laid white eggs and did not have crests.

Professor Punnett had some Araucanas at Cambridge. They were still very much a novelty back in the 1940s. Not much was known about them, but there were some in the UK. There was an article about them in the 1935 *Feathered World Yearbook*, and a major feature in the September 1948 issue of *National Geographic* magazine. Most of Professor Punnett's Araucanas were Black-Red/Partridge, but there were some Creams among them. These Araucanas were tailed, had crests and, of course laid blue eggs. They did not carry the barring gene, and did not Autosex-link.

A test mating was tried between Cream Araucanas and 'Mark I' Cream Legbars, mostly out of curiosity about the genetics of the cream plumage. Eventually a separate strain of blue egg-laying Cream Legbars was established. It was decided to retain the crests, a modest tassel of feathers behind their single combs, to avoid confusion between them and Gold or Silver Legbars. The new Cream/Crested Legbars were first displayed by the Autosexing Breeds Association at the 1947 London Dairy Show, and eventually standardized by the PCGB in May 1958.

In the 1940s and 1950s the British public was not ready for blue-shelled eggs, they were very conservative in their eating habits, and remained so until foreign holidays became cheap enough for the majority of the population. The largest laying flock of Crested Legbars at this time (1952) was about 150 hens owned by Mr Archie McKim of Lamboum, Berkshire. Blue eggs are big business now, with several commercial flocks based on mixtures of Araucanas and Cream/ Crested Legbars, probably crossed with hybrids.

Description

Body shape and size, apart from the crest, is as Gold and Silver Legbars, although some have been said to be slightly bigger and more docile, from their Araucana ancestors. Male neck, back and saddle feathering is mostly creamy, with some black centre striping broken into 'dashes' by the barring gene. Breast, leg and tail feathering is fuzzily barred in shades of grey. Females have cream necks, with a little black striping, salmon breasts and the back and wings are softly barred in shades of grey.

Crested Legbar Bantams have not been produced, but may be bred in the future if the large version continues its current increase in popularity.

DORBAR

Dorbars were an early (circa 1941) attempt by the Cambridge team to produce a heavy Autosexing Breed for meat production. Silver Dorbars were made from Silver Grey Dorkings and Barred Rocks, Gold Dorbars probably came from Red Dorkiings and Barred Rocks, but this is not certain. Dorkings had essentially been show birds since the First World War, and although there were some very large specimens around in 1941, they had not been selected for food conversion rate, growth rate or egg production for several decades, being essentially show birds in the hands of Mr A.J. Major of Langley, Bucks and a handful of others. The resulting Dorbars were more productive than Dorkings, but were never going to be a practical table breed. One wonders why they made them in the first place.

Gordon S. Hart (ABA Council Member) wrote an article about Dorbars in the 1949 *Autosexing Annual* that concluded: 'We [at Cambridge] have finally decided that we shall discontinue this breed at the end of the season.' By 1949 there were already more Autosexing Breeds, some of which had been developed elsewhere (*see* below), and the Cambridge team could only do so much. Legbars were obviously their best layers, Cambars were thought to still have potential as a dual-purpose breed, and Brussbars were better table birds.

They still had a few Dorbars left at the Cambridge Poultry Research Unit in 1957 when Michael Pease retired as its Director, and had to clear out all the Autosexing Breeds as well as clear his desk. Poultry research was clearly going to move on 'under new management'. It is not known exactly when Dorbars finally died out, but it was probably about 1960. There were never any Dorbar Bantams, and it is unlikely we shall ever see Dorbars again in either size.

Description

Dorbars were fully described in the 1954, 1960 and 1971 editions of British Poultry Standards. They have been dropped from more recent editions because they are believed to be extinct, but the previous standard would still apply if they were remade.

Both the written description and standard weights confirm a smaller and more active type than pure Dorkings. Weights ranged from 5lb (2.25kg) pullets, up to 8lb (3.6kg) adult cocks. This was never going to be large enough for a serious table breed.

Only single combs were allowed on Dorbars, unlike Dorkings where rose combs are also permitted on some colour varieties. Dorbars were required to have five toes, as Dorkings, but differed in having much tighter plumage. Plumage colour and markings of Gold and Silver Dorbars were similar to Gold and Silver Legbars.

WELBAR

Welbars are the Autosexing version of the dark brown egg laying Welsummer, and along with blue egg laying Crested Legbars, are likely to be the most popular Autosexing Breeds in the future. Welbars were not made by the Cambridge team, but by Mr H.R.S. Humphreys of Eastwrey, Lustleigh, Devon. He thought of the idea in 1940, and contacted Professor Punnett for advice. Following the Legbar process, chicks from the initial Welsummer × Barred Rock were hatched in 1941, then cockerels from this mating were bred with Welsummer hens. Gold and Silver Welbars were developed simultaneously.

There was a fair amount of variation in size, shape and colour for some years among the prototype Welbars. Mr Humphreys decided right from the start that his ideal size, shape and eggshell colour would be exactly as Welsummers. Naturally, the initial Barred Rock hens used affected egg colour, and if his new breed was to be as productive as Mr Humphries hoped, there was bound to be some falling off in shell colour. A hen's body can only produce so much pigment. It is quite normal for Welsummer eggs to be lighter-coloured later in the laying season than their first pullet eggs.

Mr Humphreys was keen to reduce the variation in plumage colours he was getting for several reasons, one being that variations in adult plumage colouring might bring differences in chick colouring, which could complicate sex identification. He wrote at length about the range of colours and patterns he bred in the 1946 *Autosexing Annual*. He was correct in assuming that most of the pattern variations came from the original Barred Rocks, but he did not have any details of the genes involved. Some of the earlier generation females had heavy black neck striping, heavy black markings on back and wings, black breast lacing and absence of salmon breast. This pattern was caused by the 'Brown, eb' gene on the 'E. locus', but the 'E series' of genes were unknown at the time.

Other early Welbars were at the other extreme – that is, too much 'red', or in the case of Silver Welbars, too much white, and not enough black. This is frequently seen in some strains of Welsummers today. Cockerels often have the standard breast colour, a mixture of black and red (ideally about two-thirds black), and by the time they are three or four years old have almost completely 'red' (orangy chestnut) breasts. Hens of this type have reduced markings on back and wings, suggesting some Rhode Island Red ancestry.

Welbar Bantams were made by John Buck of Bristol, also remembered for his excellent Brahma Bantams. They are (2005) very rare, probably less than five flocks, but have the potential, again because of the dark brown eggs, to become popular among

Welbar, bantam female. Photo: John Tarren

bantam keepers more interested in eggs than showing. They can be quickly multiplied by Welsummer crosses.

Description

The PCGB Council approved the standard for Welbars in October 1948. Curiously, only the Silver plumage variety is described. Only Golds, and now also in Bantam form, have been seen in recent years. Size, shape, and so on, is all as for Welsummers.

There is a Silver Welsummer, but very few are seen. It is a dark version of the Silver Duckwing pattern as seen on Old English Game and many other breeds. Males have a mix of black and white, usually creamy-white, breast feathering. Female Silver Welsummers have a very rich salmon-red breast colour. Silver Welbars are obviously of the same basic pattern broken and lightened by barring.

Gold Welbars are essentially as Creles in other breeds. Like all Autosexing males, only light-coloured males, with two barring genes are kept, instead of the darker (one barring gene) usually shown in other breeds. Gold male Welbar breast plumage is very variable.

RHODEBAR

As Rhode Island Reds were arguably the most commercially important chicken breed of the 1940s and 1950s, it is not surprising that Rhodebars were made, and made independently by several teams. Some were initially named 'Redbars', including the first strain detailed below.

A team (E.A. Lloyd, A.T. Hill, Ursula Knight) at the University of British Columbia, Vancouver, Canada, started in the early 1940s with a Barred Rock (Canadian utility strain) F × Rhode Island Red (also a Canadian utility strain) M/M mating. The best of the dark barred cockerels were mated to two pens of Rhode Island Red hens, one group being their mothers. These cockerels were retained for the next season, when they were mated with the clearest red-barred daughters bred from them. This mating produced a very variable third generation that included dark barred males and non-barred females that were discarded, and light-barred males and barred females that were closely examined for further breeding. They were looking for clear barring over a red ground colour, with black only appearing where it would on Rhode Island Reds. Because of the variability in plumage colour in the early stages, accuracy of sexing was well below the level needed for commercial use. In 1945, tests on sexing ranged from 60 to 80 per cent accurate, improving to ±80 per cent for all batches tested in 1946. Careful selection of breeding stock brought rapid improvements, averaging 93.6 per cent accurate in 1947. By 1950, it was quite normal to get 98 per cent of the chicks correctly sexed, sometimes 100 per cent right. Redbar hens laid at least 200 eggs per year, some nearer 250. The strains of Rhode Island Reds and Barred Rocks used were rapid feathering with tighter coloured plumage than exhibition birds. Show Rhode Island Red breeders look for the red colour to extend right down each feather to the skin, and Barred Rocks in the show halt have crisp barring down to the skin. In both cases this gives dark feather stubs on plucked birds, the last thing commercial table chickens want. Utility Rhodes and Barred Rocks have rapid feathering and light under-colour.

In 1946, Mr Folsach and Gordon Hart, of Harlow, Essex, obtained some hatching eggs of a Danish strain of Rhode Island Reds. This strain was known for being very uniform in colour, which they hoped would reduce variability problems later. When reared, in 1947, Danish Rhode pullets were mated to a Gold Brussbar (*see below*) cockerel. The resulting chickens were a uniform lot, dark barred, grey and red with white shanks (as Brussbars). They were mated together, brother × sister. This gave the expected very variable second generation predicted. The double-barred cockerels and barred pullets were retained, and the single-barred cockerels and non-barred pullets discarded. Later generations were selected for Rhode characteristics such as yellow skin/shanks/feet and restriction of black pigment to tail and wings, plus perhaps a little neck striping. These were also called 'Redbars'.

Mr B. de H. Pickard of Storrington, Sussex made his 'Rhodebars' from Rhode Island Reds and Barred Rocks, similar to the Canadian team. It is believed that it was birds of his strain which were shown to the PCGB Council in October 1951 for standardization, which after due process, was approved in October 1952. Mr Pickard entered groups of pullets in several egg laying trials in the 1950s, and he was still breeding them in 1964.

There are (in 2005) still a few flocks of Rhodebars, plus Rhodebar Bantams, the leading breeder of both being Brian Sands of Lincolnshire, also a leading figure in the revival of the Lincolnshire Buff.

Description

General body shape and size of both large and bantam Rhodebars should be the same as for Rhode Island Reds, but they seldom have quite as much length of back. Most plumage is barred, red and white, with black and white barred tails, wing markings and hackle

Rhodebar, large male. Bred by Brian Sands of Lincolnshire. Photo: John Tarren

striping. Males usually have sharper and clearer barring than females.

HAMPBAR

The Canadian team mentioned above crossed some of their Redbars/Rhodebars with New Hampshire Reds, then a popular commercial breed in Canada and the USA. The university team only bred Hampbars for a few years, circa 1948–52, then passing them on to Mr George Routledge, of New Westminster, British Columbia. They probably died out during the 1960s and no records of them in the UK have been found. Hampbars were never reliably autosexing, with a lot of 'in between' colour chicks.

BRUSSBAR

Utility strains of Light Sussex were a major utility variety in Britain from about 1920 until pure breeds were replaced by hybrids in the 1960s, so the Cambridge team obviously would have wanted to make an Autosexing Breed based on them But they could not see how to do it from a Columbian pattern bird. They decided to get around this problem by crossing some utility Light Susex with exhibition Brown Sussex.

The author has not been able to discover exactly when or how this programme started, but Punnett and Pease started publishing information about Brussbars in 1941/2. They knew before they started that a Brussbar made from just Barred Rocks and Brown Sussex would be a failure. Only exhibition strains of Brown Sussex have ever existed, and they were unlikely to produce very good layers or quick-growing table birds. As they could not think of any other method, they started with the Barred Rocks and Brown Sussex, and once they had made initial Brussbars (that were not brilliant producers), they crossed utility strain Light Sussex cocks with Brussbar hens, mated the resulting unbarred Columbian pullets with Brussbar cockerels, and selected the partridge patterned chicks from there.

Mr C.E. Marshall of Pikes Farm, Haversham, Bucks made his own strain of Brussbars, choosing an unconvential route. A Gold Dorbar cock was mated with a single selected Barred Rock hen, and from these a barred cockerel was kept for the next stage. At the same time as the first mating, another Dorbar cock was mated with Light Sussex hens. This mating produced some buff (barred) Columbian coloured pullets that were mated with the barred cockerel. This mating gave a wide range of colours in the next generation, enough to provide the foundation of both Gold and Silver 'Brussbars'. In case any readers did not notice, Mr Marshall's 'Brussbars' did not involve any Brown Sussex.

Brussbars, from whatever source, were considered good meat and egg producers by the Autosexing enthusiasts, but not by many other people. They were standardized by the PCGB in October 1952, and although they never attracted much interest, have just managed to survive to the present day. Eleven breeders were listed in the 1959 *Autosexing Annual*. By 1969 there was probably only one breeder, F.E.B. Johnson, proprietor of the long since closed-down Stagsden Bird Gardens near Bedford (and not very far from Haversham). The author remembers seeing them there, and probably should have bought some. Mr Johnson sold the 'Bird Gardens' (it was open to the public) and he may have taken some Brussbars with him when he retired. The new owners kept them going until the collection was sold off and the Gardens closed down following repeated thefts of valuable parrots and macaws. Andrew Sheppy (Rare Poultry Society founder) also had a few that he just about managed to keep going to the present (2005), as part of a collection of many rare breeds. Andrew's small flock of Brussbars is probably the total world population at time of writing.

Description

The Brussbar Standard description of size and shape is very similar to the Sussex Standard. The printed standard has never attempted to accurately describe the differences between exhibition and utility types of Light Sussex. Show birds look a lot bigger than utility types, especially those strains that had been developed for eggs rather than meat, however some of the difference is due to feather quantity and texture rather than body weight. Minimum weights of Brussbars

Brussbar male with two crossbred females. One of Andrew Sheppy's few known birds. Photo: David Scrivener

were given as: cock 9lb (4.1kg), cockerel 8lb (3.6kg), hen 7lb (3.2kg), pullet 6lb (2.7kg).

Brown Sussex are a very dark version of the Black-red/Partridge plumage pattern. For example, Brown Sussex males have 'rich dark mahogany striped with black' neck and saddle hackles instead of the bright orange-red shades of Brown Leghorns. Thus Gold Brussbars are a lot darker than Gold Legbars. It is surprising really as Brussbar males still have two barring genes, like all the other Autosexing Breed males, and one would have thought that the Light Sussex crosses used in their development would have had some effect on the final colour. Gold Brussbar males (with two barring genes) are much the same general colour as exhibition crele males (with one barring gene) of other breeds, a mixture of rich gold and straw with black centre striping broken by the barring gene. Breast, leg and tail plumage is barred, dark and light grey, ideally without gold shading. Females have dark gold neck hackles with broken black centre striping; rich salmon breast; softly barred grey and gold back and wings, with gold shafts and edging.

Silver Brussbar males (now extinct) had the same pattern, with silver replacing gold, although most were said to have had some chestnut colouring. Silver females should have had softly barred, light and dark grey, back and wings, with silvery shafts and edging.

BROCKBAR

'Brock' is an abbreviation of Buff Rock, so Brockbars were made from crosses between Barred and Buff Plymouth Rocks. Size and shape were therefore as for Plymouth Rocks, but of the utility type, not the very tall and tight-feathered British exhibition type. The plumage colour and pattern they were aiming for was distinct barring all over of orange and pale lemon.

Brockbars appeared about 1946, but seem to have died out sometime during the 1950s. There were two reasons why Brockbars failed: (1) They were not good enough as either egg producers or table birds. (2) The light plumage colour meant that the chicks of both sexes were fairly light, there were always a lot of difficult-to-sex, mid-colour chicks, which was also a problem with Hampbars and Buffbars.

'Golden Barred Plymouth Rocks' had briefly appeared, circa 1925–30, but if anyone had ever noticed that the chicks were sexable, they do not seem to have made it public. 'Golden Barreds' were mainly promoted by a Mr Whitcombe of Frome, Somerset. This strain was of the tall, British exhibition Rock type. Both Golden Barred Plymouth Rocks and Brockbars are extinct.

Both breeds, or at least the best of them, were very attractive. The orange and white (or nearly so) barring was only really sharp and clear on male neck and saddle hackle feathers. Normal body feathering is very indistinctly barred; it can hardly be seen at all on older birds, especially if they have been out in the sun.

BUFFBAR

This Buff Orpington-based Autosexing Breed had the same problem as Brockbars, about 10 per cent of the chicks were difficult or impossible to sex. It is believed that they were first made about 1945, died out during the 1950s, and were never standardized by the PCGB Only one flock was known by 1952, that of Mr J.S. Freckleton of Ashby-de-la-Zouch, Leicestershire.

ANCOBAR

This Ancona-based Autosexing Breed was first made by (Professor) Lamoureux in the US in 1941. A Mr P.A. Francis of the British Ministry of Agriculture (then 'MAFF', now 'DEFRA') suggested that the Cambridge team make a British version of the same. Ancobars were a low priority at Cambridge however, and it was not until 1956 that they had any quantity of them. This was only shortly before Michael Pease retired from Cambridge, Professor Punnett having long since retired. With both of the Autosexing enthusiasts gone from Cambridge, the Ancobar was never going anywhere.

Description

Shape, size, comb, lobes, and so on were all as Anconas. There was considerable difference in adult plumage colour and markings between males and females. Males were mostly white with cuckoo tails and irregular small black spots over the remainder. Ancobar females were much darker, cuckoo barred all over, plus a sprinkling of white feather tips as one would expect from an Ancona-based breed. This irregular plumage pattern was never going to appeal to exhibitors.

WYBAR

Lt Cmdr A.O. Foden spent thirty years in the poultry industry after he left the Royal Navy. For part of that time he worked for Michael Dewar (one of the Whisky distilling Dewars) at his estate near Bletchley, Buckinghamshire. One of the breeds kept there was the Legbar, which led to Lt Cmdr Foden taking an interest in Autosexing Breeds generally. When Lt Cmdr Foden retired, he moved to Hadleigh, Essex, and started to keep his own poultry rather than that of other people. He had always liked Wyandottes, so decided to make a Wyandotte-based Autosexing Breed.

His first Wybar chicks were hatched in 1950 or 1951, which suggests that breeding operations started about 1947 or 1948. He imagined they would be a successful 'triple-purpose' breed, good for eggs, meat and exhibition. They never succeeded in any of these functions. As far as egg production was concerned, Wybars laid an average of 230 in their first year, with 'many' as high as 260. This was fairly good, but nothing exceptional, for that period. Wyandottes have had a reputation for small egg size. It is not known if Wybars were any better.

Gold and Silver Wybars were made using Gold and Silver Laced Wyandottes, Barred Rocks and Brussbars. Remembering that Gold and Silver Laced Wyandottes in the UK then were taller and tighter-feathered than White and some other colours of Wyandottes, and considering the other breeds involved, it is not surprising that Wybars were not as full-feathered and rounded as some, especially American readers, might expect.

Lt Cmdr Foden and John Croome still had Wybars in 1964. It is not known how much longer they persevered with them. It is doubtful if there were any left after 1970. Silver Wybar Bantams have been seen at the larger shows since about 2000, the main exhibitor being Mrs J. Ashton of Sheffield. It is not known if she made them.

Description

A standard for Wybars was adopted by the PCGB on 9 February 1956. The description of size and shape was very similar to the existing Wyandotte standard, but as said above, they were not really like that.

The wording of the plumage patterns in the published standard was not very clear. The photos in the *Autosexing Annual* help somewhat, but are small and grainy, so are not ideal. If you begin by imagining normal Gold and Silver Wyandottes, and then remember that Wybars were supposed to be the same, but modified by the barring gene, it is fairly easy to work out what ideal Gold and Silver Wybars should look like. The general impression is of light-coloured Laced Wyandottes with cuckoo-barred tails. They should not be confused with Crete Wyandottes, the barred variant of the Partridge pattern.

WHEALBAR

'Wheal' means 'mine' in the old Cornish language, which suggests that Mr E.W. Tresidder, of Padstow, Cornwall, hoped his 'Whealbars' would be a gold mine. Mr Tresidder gave a few details in the 1950 *Autosexing Annual*, but not a full account of his matings. Perhaps

Silver Wybar, bantam male. Photo: John Tarren

Silver Wybar, bantam female. Photo: John Tarren

he thought others would copy his wonderful new breed. The main breeds used may have been Red Wyandottes, Rhode Island Reds, (Light?) Sussex and an unspecified Autosexing Breed, probably either Rhodebar or Brussbar. The resulting Whealbars had red-barred plumage (similar to Rhodebars), rose combs, white skin/shanks/feet, and were intended to be a hardy dual-purpose breed. Mr Tresidder traded as 'Grafton Poultry Industries Ltd', and considering little, if anything, was ever heard of Whealbars again, having limited company protection was probably a wise precaution.

OKLABAR

One of the important American poultry experts was Dr R. George Jaap of the Oklahoma Agricultural Experiment Station. He followed the work of Punnett and Pease at Cambridge on the Autosexing Breeds and tried to incorporate this principle in strains he was developing for the rapidly growing American poultry industry of the 1930s and 1940s. His work was continued through to the early 1950s by Dr George F. Godfrey at the OAES, Stillwater, Oklahoma.

They started with crosses of Rhode Island Reds and a strain of White Plymouth Rocks (then a very important commercial breed in the USA) that was discovered to be carrying the barring gene. These crosses were then mated with 'Dark Cornish', the American poultry meat industry's version of the UK's Dark Indian Game. Progeny from these three-way crosses were selected for reliable Autosexing and became 'Gold Oklabars'. Some of these were crossed with some Silver Laced Wyandottes to make Silver Oklabars. The Silvers were never reliable as Autosexers, there not being enough difference in colour between the sexes.

The ability to identify the sexes of meat chicks became less and less important as the beginning of today's broiler industry developed. Fast-growing chickens with white feathering, so there were no coloured stubs, were the only acceptable type. Oklabars have never been a fanciers' breed and they probably became extinct before 1960.

Autosexing Breeds

NIEDERRHEINER

See main entry in European Heavy Breeds chapter. One of the colour varieties is 'Kennfarbig', light Crele, as with Welbars in the UK. The 'Gelbsperber' variety, similar to Brockbars, may produce visually sexable chicks, but the author was not certain of this at the time of writing.

BIELEFELDER KENNHUHN

This German Autosexing Breed was developed by Herr Gerd Roth of Bielefeld, a large town halfway between Dortmund and Hanover. His new breed, first called 'Deutsches Kennhuhn', was standardized in Germany, under its new name, in 1980, with the bantam version following in 1985. Large Bielefelders were made from Cuckoo Malines and

Bielefelder, large male. Photo: Hans Schippers

Bielefelder, large female. Photo: Hans Schippers

Welsummers. Undersized large Bielefelders were crossed with Amrock and Welsummer Bantams to make the 'Bielefelder Zwerg-Kennhuhn'. Although a new breed, Bielefelder Kennhuhner of both sizes have become very popular, as indicated by entries of well over a hundred birds regularly appearing at German shows. Because they are similar to Welbars, it is unlikely that Bielefelders will catch on in the UK, but it does suggest that Welbars have a lot of potential over the next few years. Bielefelders lay dark brown eggs, often with very dark spots.

Description

As expected from a Welsummer-based breed, they have medium-sized single combs, red lobes, red eyes and yellow skin/shanks/feet. Carriage is more horizontal than Welsummers or Welbars because Bielefelders were also made from heavy and docile Cuckoo Malines. Tail carriage is only about 30 degrees above the horizontal. Plumage is quite thick and profuse, about the same as exhibition-type Light Sussex.

Bielefelders are very similar in plumage pattern to Gold Legbars and Gold Welbars. Because of the Welsummer ancestry, mates are somewhat variable in breast colour, some are nearly all red and white, others have more grey and white. The females are more uniform, with salmon breasts, gold/black/white neck hackles and brown/grey marbled back and wings.

Standard weights: large male 3–4kg (6½–8¾lb), large female 2.5–3.25kg (5½–7lb). Bantam male 1200g (42oz), bantam female 1,000g (35oz).

NORSKE JAERHØN

This Norwegian breed is described in the European Light Breeds chapter because although both colour varieties are autosexing, it is a traditional old breed, not the product of scientists and their followers in the twentieth century, like the rest in this chapter.

POLBAR

These were developed around 1946 by Professor Kaufmann in Poland from utility type Barred Plymouth Rocks and other unrecorded breeds. They are a light cuckoo barred breed, with the males having very silvery neck, back and saddle feathering. Polbars are unlikely to be of interest to hobbyist poultry keepers and are only included in this book to complete the list of known Autosexing Breeds.

Bibliography

All published in UK unless otherwise stated.

American Bantam Association, *Bantam Standard*, USA, 1979. There are other editions

American Bantam Association, *Book of Bantams*, USA, 1967

American Bantam Association, *Yearbooks*, USA, 1980, 1981

American Poultry Association, *The American Standard of Perfection*, USA, 1890, 1974, 1985

American Poultry Journal, 1913 Yearbook, USA

Avicultura magazine, The Netherlands, 1980–2001

Autosexing Breeds Association, *Autosexing Annual*, 1946–64

Bantam Club of New South Wales Yearbooks, Australia, 1980, 1981–82

Bennett, Dr John C., *The Poultry Book*, USA, 1850

British United Bresse Club Yearbooks, 1928, 1933

Broomhead, William W., *Poultry Breeding and Management*, circa 1930

Brown, Edward, *Races of Domestic Poultry*, 1906 (reprinted 1985)

Brown, Edward, *Poultry Breeding and Production*, 1929 (became Sir Edward Brown in 1930)

Brown, J.T., *The Encyclopaedia of Poultry*, 2 vols, 1909 and revised by J.S. Hicks edition, 1921

Browne, D.J., *The American Poultry Yard*, USA, 1857

Catalogues of numerous poultry shows, UK, USA, Belgium, Germany, Netherlands

Country Smallholding magazine, 2000–2005

Der Grosse Geflügelstandard, vols 1 and 2, Germany, circa 1980

Doyle, Martin, *The Illustrated Book of Domestic Poultry*, 1857

Easom Smith, Harold, articles in *Poultry World* magazine, 1968 to 1974

Easom Smith, Harold, *Bantams for Everyone*, 1967

Easom Smith, Harold, *Modern Poultry Development*, 1976

Entwisle, William Flamank, *Bantams*, 1894 (reprinted 1981)

Fancy Fowl magazine, vol.1, no.1, October 1981 to date

Feathered World magazine, various issues, 1892 to 2004

Feathered World Yearbooks, 1910 to 1938

Fields, Mark A., *The American Dominique*, USA, 1997

Geflügel-Börse magazine, Germany, 1969–95

Houdan and Crève-Coeur Breeders' Association Newsletters, 1954–57

House, Charles Arthur, *Bantams and How to Keep Them*, circa 1930

Houwink, R. (text) and Theo Dijkwel (pictures), *Onze Hoenders*, The Netherlands, 1929. (This was a set of 144 cards, with album, given with Johan van Oldenbamevelt Cigars)

Jeffrey, Fred. P., *Bantam Breeding and Genetics*, 1977 (other editions 'Bantam Chickens')

Jull, M.A. *et al.*, *The Races of Domestic Fowl* and other articles in *National Geographic* magazine, April 1927 issue, USA

Bibliography

Kay, Ian, *Stairway to the Breeds*, 1997

Keeling, Julia, *The Spirit of Japanese Game*, 2003

Kramer, Rudolf, *Tacshenbuch der Rassegeflügelzucht*, Germany, 1926 (reprinted 1999)

Lamon, Harry M. and Rob R. Slocum, *The Mating and Breeding of Poultry*, USA, 1927

Lee, Charies, *The Houdan Fowl*, 1874

Lewis Jones, Rev., E., *The Campine*, circa 1911

Lewis Jones, Rev. E., *The Production of the English Type Gold Campine*, 1914

Lind, LR., *Aldrovandi on Chickens*, USA, 1963 (translation from Latin of the book published in 1600)

Long, James, *Poultry for Prizes and Profit*, circa 1890

McGrew, T.F., *The Bantam Fowl*, USA, 1899 (reprinted 1991)

Moubray, Bonington, *Domestic Poultry*, 1842

Nederiandse Bond van Hoender-, *Dwerghoender, Sier-, Watervogel, Hoenderstandaard*, 1986

New Zealand Poultry & Pigeon News Digest, NZ, 1999–2002

Ogasawara, Dr Frank X., 'Japan's Fantastic Long-tailed Fowl', *National Geographic* magazine, December 1970 issue, USA

Pigeons & Bantams magazine, vol. 1, May 1947–April 1948

Poultry Club of Great Britain, *The Poultry Club Standards*, editions from 1886 to 1926

Poultry Club of Great Britain, *British Poultry Standards*, 1954, 1960, 1971, 1982, 1997

Poultry Club of Great Britain, *Yearbooks*, 1927 and 1965 to 2004

Poultry magazine, bound volume of 1901 issues

Poultry magazine, 1923 and 1924 Yearbooks

Poultry Press magazine, USA, 1980–2000

Poultry World magazine, various issues 1918 to 2005

Poultry World Yearbooks, 1914, 1917, 1920, 1930

Practical Poultry magazine, no. 1, April 2004–no. 22, January 2006

Proud, Pringle, *Bantams as a Hobby*, 4th edition, 1912

Rare Poultry Society, *Newsletters, 1971–2004*

Roberts, Victoria, *Poultry for Anyone*, 1998

Robinson, John H., *Popular Breeds of Domestic Poultry*, USA, 1924

Schmidt, Horst, *Handbuch der Nutz- und Rassehühner*, Germany, 1985

Shakespeare, Joseph, *The Bantams Down-to-Date*, USA, 1925

Show Bird Journal, USA, 1998–2000

Silk, W.H., *Bantams and Miniature Fowl*, 1951, 1974

Smith, Page and Charles Daniel, *The Chicken Book*, USA, 1975

Somes, Ralph G., *Registry of Poultry Genetic Stocks in North America*, USA, circa 1975

Stromberg, Loyl, *Poultry Oddities, History, Folklore, USA*, 1992

Various contributors, *Federfüssige Zwerghühner*, Germany, 1987

Verhoff, Esther, and Aad Rijs, *The Complete Encylopedia of Chickens*, The Netherlands, 2003

Wandelt, Rudiger and Josef Wolters, *Handbuch der Hühnerrassen*, Germany, 1996

Wandelt, Rudiger and Josef Wolters, *Handbuch der Zwerghuhnrassen*, Germany, 1998

Watts, Elizabeth, *The Poultry Yard*, circa 1870

Weir, Harrison William, *The Poultry Book*, USA edition, 1905 (1st UK edition, 1902)

Willems, Prof. A.E.R. and E.T. Brandt, *Studie over Hoenderachtigen en Watervogels*, Belgium, 1971

Wingfield, Rev. W.W. and G. Johnson, *The Poultry Book*, 1853

Woods, Rex, *Rare Poultry of Ashe*, 1976

Wright, Lewis, *The Book of Poultry, Popular Edition*, 1888

Wright, Lewis, *The New Book of Poultry*, 1902

Wright's Book of Poultry, revised by S.H. Lewer *et al.*, 1919 (after Wright's death)

Wulfften Palthe, A.W., *C.S. Th. van Gink's Poultry Paintings*, The Netherlands, 1992

INDEX

Adam, Paul 35
Albion (US) *231*
Aldrich, Dr., N.B. 232
Aldrovandi, Ulisse 107, 164, 235
Alsacienne *10*, 48, 58
Altsteirer *123-5*
American Seabright *231*
Ancobar *261-2*
Andalusian *10–14*
Annaburger Haubenstrupphuhn *163*, 168
Antwerp Brahma 201, 209
Appenzeller Barthuhner *14*
Appenzeller Spitzhauben *125,* 147, 152, 153
Appleyard, Reginald 205–6
Arbeiter, Amin 148
Ardennaise/Ardenner Hoen *14–16,* 39
Ardennaise Sans-Queue/Ardenner Bolstaart (Rumpless Ardennes) 15, *108,* 109, 110, 112
Assendelftse Hoen *16*, 53, 72
Aßmus, L 110
Audinger, L.B. 155
Augsburger 145, 160–1
Australian Langshan 193

Baerman, Freeman Donald 139
Baker, W.H. 33
Balady 79
Barber, Leonard 10
Barbezieux *200*
Barbu d'Everberg *108–9*
Barbu du Grubbe *109*
Bateson, William 251
Bechstein, J.M. 73
Belgische Kriel (Belgian Bantam) *235–6*

Bennett, Dr. John C. 133, 153, 164, 220, 224, 232
Bergischer Kräher *115–17*
Bergischer Schlotterkamm *16–18*
Beyrich, Arno 104
Bicknell, J.Y. 224
Bigawi *80*
Bielefelder Kennhuhn *264*,
Billson, Richard 104, 151, 158, 190, 247
Birmingham White *231*
Black Brothers, John & Thomas 227
Booted Bantam *235–43*
Borgards, Heinrich 202
Bourbonnaise *200–1*
Bourbourg *201*
Brabançonne/Brabants Boerenhoen *128–30*
Brabanter *130–L* 143, 152
Bracken, Thomas B. 23–4
Brake! 17–21, 30, 39, 53, 209
Branwhite, F.C. 153,
Brassington, Syd. 33
Breda *131–4*, 143
Bresse-Gauloise *20–3*, 29, 48, 212
Bristol County Fowl *231*
Bristol White *231*
Brockbar *261*
Brown, Sir Edward 14–15, 18, 22, 23, 27, 28, 29, 35, 44, 45, 53, 73, 77, 89, 111, 134, 143, 148, 200, 202, 209, 243
Brussbar *259–61*
Buckeye *226–7*
Buck, John 257
Bucks County Fowl 220, *231*
Buffbar *261*
Burmese Bantam *243–4*

Burney, Dr. 152–3
Butters, J.A.C. 97

Cambar 252–3
Campbell, Mrs Adeie 151
Camphausen, Ewald 59
Campine 18, *23–7*, 30, 53, 209, 252–3
Carswell, John 61–2
Casteliana Negra/Castilian 27–8
Castelló, Prof. Salvador 28
Catalana del Prat 28–9, 38
Caumont *134*,148
Caussade *29*, 48
Caux 29
Ceská Slepice *29–30*
ChaamseHoen 17, *30*
Chantecler 227
Chateaine, Brother Wilfred 227
Chatterton, F. J. S. 62, 93
Chilianskaia (Oriofť) *89*
Chinese Langshan 193
Cotentine *30*
Coucou de Bretagne *202*
Coucou de Flandres *201*, 209
Coucou de France *202*
Coucou d'Iseghem *201–2*
Coucou de Rennes *202*
Courtes Pattes *175*
Coveney White *159–60*
Craig, J. 178
Cream or Crested Legbar *255–6*
Cremat, Hauptmann Karl 202
Cresswell, Oliver E. 245–6
Crève-Coeur 134, 135–7, 138, 142
Croad Langshan 181,184-8,204
Croad, Major F.T. & Miss A.C. 181, 183–4,186, 188, 189, 191, 193, 204
Croome, John 252, 262
Cubalaya *83–4*

Dansk Landhene/Danish Landhen *31*
Danvers White *231*
Dartmouth, Countess 13
Darwin, Charles 65, 108, 183
deLame.Théo 108
Delaware *230–1*
Delin, Rene 202
Denizli *117*

Derbyshire Redcap *31–5*
D'Estaires *204*
Detroy, A. 201
Deutscher Sperber *34–6*
Deutsches Langschan *193–7*
Deutsches Reichshuhn *202–3*
Deutsches Zwerghuhn 35, 145, *244–5*
Devereux, H.L. 133
Dillsbury, Arnold G. 25
Dominique *220–3*
Doornikse Kriel *248–9*
Dorbar *256–6*
Doyle, Martin 65–6
Drentse Bolstaart 36, *109–10*
Drentse Hoen 29, *35–7*. 39
Dresdner *203–4*
Drexter, Michael 169–70
du Roi Hugo 92, 103, 240
Dutch Owlbeard 73, 77, 131, 134, 143, *145-8*

Easom Smith, Harold 51, 61, 187
Edwards, Richard 24
Ellis, George 230
Elworthy.J.E. 153
Empordanesa *38–9*, 55
Entwisle, William Flamank 108, 141, 237, 243–4, 246
Esche, Arthur 59
Estaires *204*
Eureka *231*
Everbergse Baardkrei *108–9*
Everleigh, Gordon 193

Fammene-Hoen 39
Fayoumi *80–1*
Federffüßige Zwerg *235–43*
Felch.l. K. 138, 232
Fentiman, Mrs. 151
Fields, Mark A. 220–2
Finsterbusch, C.A. 84
Fisher, Kurt 14, 125, 145
Fitzwilliam, Hon. C.W. 143
Flemer, Ehme 53,125
Fletcher-Houseman, R, 188
Florschütz, Ernst 73
Foden, Lt. Cmdr. A.O. 262
Fraser, John C. 170
Freckleton, J.S. 261

Index

Fries Hoen/Friesian 39–42, 43, 53, 147
Frizzle 163–8

Gatinaise *204*
Gauloise Doree 20, *42*
Gebigke, Fritz 116
Gele van den Mehaigne *245*
Geline De Touraine *245*
Gemsjager, Kurt 137
German Langshan *193–7*
Godfrey, Dr. George F. 263
Golden Buff *232*
Gough, John B. 138
Gournay 43–4
Groen, L 53
Groninger Meeuwe *43*, 53, 54
Grubbese Baardkiel 109
Guelderlander *131–4*

Hagedoorn, Dr. A.L 251
Hahn, Paul 239
Hambletonian 232
Hampbar *259*
Hams, Fred 67, 70
Hart, Gordon, S. 256, 259
Hartung, Gustav 240
Heathcote, Joseph 31
Heaton, Frederick R. 97
Heermann, Rektor Johann 194
Henwood, George 135, 144
Hergnies 17, 44
Hering, Leonora 139
Herregodts, M. 45
Herve-Hoen *45*
Hickey, William 65–6
Hill.A.T. 258
Hoffmann, Dr. Edmund 230
Hoffmann, Martin 194
Holland (US) *230*
Houdan 135, *138–42*
House, C.A. 70
Houwink, Roel 36
Humphreys, H.R.S. 257
Hünlich, Kurt 170
Huth.Kari 110,169

Isherwood.C. 144,157–8
Ixworth *205–6*

Izegemse Koekoek *201–2*

Jaap, Dr. R. George 263
Jackson, Hon. Pamela 125
Jacobus, R. 24
Java 220–1, *223–5*
Jersey Blue 220, *232*
Jersey Giant *227–9*
Jitokko *175-6*
Jobs, Jaques 211
Johnson, F.E.B. 151, 260
Jonientz, Georg 163
Jurlowskije Golosistye 117

Kaufrnann, Prof. 265
Kaulhuhn 108, 109, *110–11*
Kay, John, Ian, Dean, & Stuart 28, 190
Keaffaber, Merle L. 84
Kellog, Bernard L 172
Kitchen, Fred. 33
Klein, E. 240
Knieper, Richard 202
Knight, Ursula 258
Knöll, H. 124
Knöpfler, Otto 161
Koeyoshi *117–18*
Konze, Hans 194
Kowert, Erich 222
Kraaikoppen *131–4*
Kraienköppe *99–101*, 132
Kruper *176*
Küppersbusch, Willhelm 116
Kurokashiwa *118–19*

Labbe, Edouard 201
LaFleche 48,134,138, *142–5*
Lakenfelder/Lakenvelder 45–7,147
Lamarche, Georges 108
Lambert, Miss L. 151
Lamona *229–30*
Lamon, Harry M. 229–30
Landaise *48*
Langshan – early imports *181–4*
Langshan – US Standard *191–3*
Lax, Dr. 144
Lee, Charles 138
Legbar *253–5*
Leipner, Emit 241

Le Mans *48*, 58
Lewis, John Spedan 98
Lewis-Jones, Rev. E. 24
Leyson, Thomas 215
Limousin *48*
Lincolnshire Buff *206–8*
Llewellyn, Rhys 89
Lloyd, E.A. 258
Long, James 61, 165
Loring, C. Carroll 155
Lowe, Cliff 70
Luttehøne *176*
Lyell, James C. 69
Lyonnaise *168*

Madagascan Naked Necked Game *168*
Malines/Mechelse Hoen 201, 202, *208–9*, 211
Malmburg, Pete 225
Mantes *209, 211*
Marhold, Otto 169
Marshall, C.E. 260
Marsh Daisy *48–50*
Martin, Fred. 246
Mattock, E.S. 24
Mayer, J. 160–1
Meloney, U.L. 228
Mensinger, Ernst 241
Metcalf, Mrs. Nettie 226
McGrew, T.F. 192, 238
Me Keen, Joseph 233
Miles, Leslie C.W. 214
Minohiki *85*, 103
Modem Langshan 183, *188–91*
Montane, Domingo 83–4
Moore, Charles 48–9
Morgan, Edward 166
Müller, Otto 240
Müller, Philip 241
Müller, Richard 169
Müller, Willheim 244
Murphy, Arthur D. 24
Myhill, Fred. W. 215

Nankin Bantam 108, *245–7*
Nederlandse Uilebaarden 73, 77, 131, 134, 143, *145–8*
Neiderrheiner *211–12*, 264
Neubert, Louis 93

Neudeck, Martin 74
Neumann, Hans Otto 241
Noack. B. 169
Noire du Berry *212–13*
Noire de Challans 212, 213
Noord Hollandse Hoen 211, 213
Norske Jaerhøn *50–1*
Norfolk Grey *215–16*
North Holland Blue *214–15*

Oana, H. 96
Ogasawara, Frank X. 85–6
Ohiki *247*
Oklabar *263*
Old English Pheasant Fowl 34, *51–3*
O-Naga-Dori *85–9*, 101, 103, 105
Oneida County Fowl *232*
Onken Willhelm 202
Orloff/Orlowskie *89–92*
Osborn, Stephen 164
Ostfriesische Mowe 43, *53–5*

Paetz, Dr. Günter 194
Parker, Eric 141, 151, 152, 190
Pauwets, Robert 108, 109
Pavilly 134, *148*
Pease, Michael 251, 255, 256, 260
Penedesenca 39, *54–6*
Peters, Mrs M.F. 141, 246
Pettipher, Joseph 66
Phoenix/Phönix *92–5*, 101, 103, 105
Pickard, B. de H. 259
Pickerill, John & Tom 190
Pictave *247–8*
Pierce, George 164
Pinkney, Daniel 138–9
Plymouth Rock (Dr. Bennett's) 220, *232*
Pohle, Fritz 222
Polbar 265
Pope, R. J. 188
Powell-Owen, W. 28
Proud, Pringle 69–70
Ptarmigan Fowl 150, *152–3*
Punnett, R.C. 251, 256, 260

Rackham, T. Farrer 164–5
Ramelsloher *56–7*, 58
Reber. U. 110

Index

Regenstein, Friedrich 211
Rehme, Heinz 35
Rheinländer 56, *57–9*, 145
Rhodebar *258–9*
Rice, Frank 111
Richards, T.V. 171–2
Robinson, John H. 13, 68, 164, 225
Roth, Gerd 264
Routledge, George 259
Rowan, R.A. 68
Royden, Dr. T.W.E. 98
Ruhlaer Zwerg-Kaulhuhn/Rumpless Booted 108, *111*
Rumpless Game 108, *111–13*

Sabelpoot *235-43*
Sachsenhuhn 56, 59–60
Sack, Hans 103–4
Sands, Brian 100, 207
Satsumadori *95–6*, 103
Schäfer, Josef 194
Schierloh, H. 141
Schlesinger, Emit 116
Schmudde, Horst W. 84
Schreiner, Elli 202
Schweizerhuhn *217*
Scots Dumpy *176-9*
Scots Grey 60–5
Seabright Cochin *233*
Shedd.Wm. E. 231
Sheppy, Andrew 151, 153, 190, 238, 243, 246, 260
Sherwood *233*
Shôkuku *96–7*
Sicilian Buttercup 39, 155–9
Sicilian Flowerbird *159*
Spanish *65–72*
Starumse Rondkamm 72
Stott, G.W. 49
Street-Porter, Stanley 160
Sulmtaler *148–50*
Sultan 150–2
Sumatra 97–9
Sundheimer *217*
Swedish Bantam *248*

Tagliaferro, Pompilio 164
Takechi, Riemon 85–6

Taylor, John 10
Tegetmeier, W.B. 108
Templeton, Joe. L 71
Terrot, R. 157–8
Thomas, S.W. & M.R. 135, 137, 139, 140
Thorniley, P.W. 46
Thüringer Barthuhn *73–4*
Tomaru 118, *119–20*
Tôtenkô *121*
Tournais 248–9
Transylvanian Naked Neck 168–73
Tresidder, E.W. 262–3
Trielhof, Otto 34
Trudgian, Rev. Ray. 117–18
Twentse Hoen 99–101, 132
Tyldesley, Fred 67

Ukrainskaja Uschanka *73–4*
Unwin, W.J. 159–60
Uzara-O 113

Van Bogaert, Dr. 108
Van Gink, C.S.Th. 37, 146
Vi!a, Prof. Roselli 38
Villa-Secca, Baron 193
Vits, Dr. Bridgit & Wolfgang 117
Vlaanderse Koekoek *201*, 209
Vogtländer *74*
Voitellier, M. 209
Von Langen, Dr. Hans Rudolf 58
Vorwerkhühner/Vorwerk *74-6*
Vorwerk, Oskar 74-5

Wallis, Jane 75
Ward, Brian 100
Ward, Kenneth 238
Watts, Miss Elizabeth 150–1
Wobor, C.E. 168
Webers.H.J. 148–9
Weir, Harrison 11, 22, 61,103, 108, 134, 135, 138, 139, 143, 152, 176–7, 207, 224, 233, 243
Welbar *257–8*
Westfälischer Totieger *76*
Whealbar *262–3*
White Birmingham *231*
White Bristol *231*
White Coronet *159–60*